Computational Intelligence in Photovoltaic Systems

Computational Intelligence in Photovoltaic Systems

Special Issue Editors

Emanuele Ogliari
Sonia Leva

MDPI • Basel • Beijing • Wuhan • Barcelona • Belgrade

Special Issue Editors
Emanuele Ogliari
Politecnico di Milano
Italy

Sonia Leva
Politecnico di Milano (Campus Bovisa)
Italy

Editorial Office
MDPI
St. Alban-Anlage 66
4052 Basel, Switzerland

This is a reprint of articles from the Special Issue published online in the open access journal *Applied Sciences* (ISSN 2076-3417) from 2017 to 2018 (available at: https://www.mdpi.com/journal/applsci/special_issues/Computational_Intelligence)

For citation purposes, cite each article independently as indicated on the article page online and as indicated below:

LastName, A.A.; LastName, B.B.; LastName, C.C. Article Title. *Journal Name* **Year**, *Article Number*, Page Range.

ISBN 978-3-03921-098-5 (Pbk)
ISBN 978-3-03921-099-2 (PDF)

© 2019 by the authors. Articles in this book are Open Access and distributed under the Creative Commons Attribution (CC BY) license, which allows users to download, copy and build upon published articles, as long as the author and publisher are properly credited, which ensures maximum dissemination and a wider impact of our publications.

The book as a whole is distributed by MDPI under the terms and conditions of the Creative Commons license CC BY-NC-ND.

Contents

About the Special Issue Editors vii

Sonia Leva and Emanuele Ogliari
Computational Intelligence in Photovoltaic Systems
Reprinted from: *Appl. Sci.* **2019**, *9*, 1826, doi:10.3390/app9091826 1

Guojiang Xiong, Jing Zhang, Xufeng Yuan, Dongyuan Shi and Yu He
Application of Symbiotic Organisms Search Algorithm for Parameter Extraction of Solar Cell Models
Reprinted from: *Appl. Sci.* **2018**, *8*, 2155, doi:10.3390/app8112155 4

Alberto Dolara, Francesco Grimaccia, Marco Mussetta, Emanuele Ogliari and Sonia Leva
An Evolutionary-Based MPPT Algorithm for Photovoltaic Systems under Dynamic Partial Shading
Reprinted from: *Appl. Sci.* **2018**, *8*, 558, doi:10.3390/app8040558 22

Mohamed Louazni, Ahmed Khouya, Khalid Amechnoue, Alessandro Gandelli, Marco Mussetta and Aurelian Crăciunescu
Metaheuristic Algorithm for Photovoltaic Parameters: Comparative Study and Prediction with a Firefly Algorithm
Reprinted from: *Appl. Sci.* **2018**, *8*, 339, doi:10.3390/app8030339 40

Alberto Dolara, Francesco Grimaccia, Sonia Leva, Marco Mussetta, Emanuele Ogliari
Comparison of Training Approaches for Photovoltaic Forecasts by Means of Machine Learning
Reprinted from: *Appl. Sci.* **2018**, *8*, 228, doi:10.3390/app8020228 62

Giovanni Petrone, Massimiliano Luna, Giuseppe La Tona, Maria Carmela Di Piazza and Giovanni Spagnuolo
Online Identification of Photovoltaic Source Parameters by Using a Genetic Algorithm
Reprinted from: *Appl. Sci.* **2018**, *8*, 0009, doi:10.3390/app8010009 78

Manel Hammami, Simone Torretti, Francesco Grimaccia and Gabriele Grandi
Thermal and Performance Analysis of a Photovoltaic Module with an Integrated Energy Storage System
Reprinted from: *Appl. Sci.* **2017**, *7*, 1107, doi:10.3390/app7111107 94

Kwangbok Jeong, Taehoon Hong, Choongwan Koo, Jeongyoon Oh, Minhyun Lee and Jimin Kim
A Prototype Design and Development of the Smart Photovoltaic System Blind Considering the Photovoltaic Panel, Tracking System, and Monitoring System
Reprinted from: *Appl. Sci.* **2017**, *7*, 1077, doi:10.3390/app7101077 109

Mian Guo, Haixiang Zang, Shengyu Gao, Tingji Chen, Jing Xiao, Lexiang Cheng, Zhinong Wei and Guoqiang Sun
Optimal Tilt Angle and Orientation of Photovoltaic Modules Using HS Algorithm in Different Climates of China
Reprinted from: *Appl. Sci.* **2017**, *7*, 1028, doi:10.3390/app7101028 127

Francesco Grimaccia, Sonia Leva, Marco Mussetta and Emanuele Ogliari
ANN Sizing Procedure for the Day-Ahead Output Power Forecast of a PV Plant
Reprinted from: *Appl. Sci.* **2017**, *7*, 622, doi:10.3390/app7060622 139

Ying-Yi Hong and Po-Sheng Yo
Novel Genetic Algorithm-Based Energy Management in a Factory Power System Considering Uncertain Photovoltaic Energies
Reprinted from: *Appl. Sci.* **2017**, *7*, 438, doi:10.3390/app7050438 . . . . **152**

About the Special Issue Editors

Emanuele Ogliari received his Ph.D. degree in Electrical Engineering from the Politecnico di Milano, Italy, in 2016, and has been serving here since 2012 as Assistant Professor with the Energy Department of Engineering, Faculty of Engineering. His current research interests include new forecasting methods by means of soft computing and performance assessment of the energy production by renewable energy systems.

Sonia Leva received her Ph.D. degree in Electrical Engineering from the Politecnico di Milano, Italy, in 2001, where she also currently serves as Full Professor of Electrical Engineering in the Department of Energy and as Director of the Solar Tech Lab and the Laboratory of Microgrids. Prof. Leva is a member of the IEEE Working Group "Distributed Resources: Modeling and Analysis" and a senior member of the IEEE Power and Energy Society.

Editorial

Computational Intelligence in Photovoltaic Systems

Sonia Leva * and Emanuele Ogliari *

Department of Energy, Politecnico di Milano, 20156 Milano, Italy
* Correspondence: sonia.leva@polimi.it (S.L.); emanuelegiovanni.ogliari@polimi.it (E.O.);
  Tel.: +39-2399-8524 (E.O.)

Received: 26 April 2019; Accepted: 29 April 2019; Published: 2 May 2019

Keywords: computational intelligence; day-ahead forecast; photovoltaics

---

Photovoltaics, among renewable energy sources (RES), has become more popular. However, in recent years, many research topics have arisen, mainly due to problems that are constantly faced in smart-grid and microgrid operations, such as output power plants production forecast, storage sizing, modeling, and control optimization of photovoltaic systems.

Computational intelligence algorithms (evolutionary optimization, neural networks, fuzzy logic, etc.) have become more and more popular as alternative approaches to conventional techniques in solving problems such as modeling, identification, optimization, availability prediction, forecasting, sizing and control of stand-alone, grid-connected, and hybrid photovoltaic systems. In this Special Issue, the most recent developments and research for solar power systems are investigated. There are ten papers selected to focus on computational intelligence methods employed in solar energy systems.

Jeong et al. [1] designed and developed prototype models of smart photovoltaic system blind (SPSB) by evaluating PV panel, tracking system, and monitoring system. This study shows that a-Si PV panel are linked in parallel, by applying four tracker types and the direct tracking method based on electricity generation with the monitoring system that can establish a time-series database on the electricity generation, the environmental conditions, and optimal tilted and azimuth angles has the best configuration.

Hammami et al. [2] conducted thermal analysis to evaluate cell temperature and battery temperature in different environmental conditions to determine the thermal limits in the 1D thermal model using the thermal library of Simulink-Matlab, with or without a set of lithium-iron-phosphate (LiFePO4) flat batteries back side of PV module. The model validation has been carried out considering the PV module to be at Normalized Operational Cell Temperature (NOCT) given by the manufacturer, and by specific experimental measurements on the real PV module, including thermographic camera images, with and without the proposed Battery energy storage systems (BESS).

Hong and Yo [3] propose an enhanced genetic algorithm (GA) to deal with unit commitment (UC) and demand response (DR) considering uncertain amounts of generated power from renewable sources in the factory power system. The uncertainty of PV power is modeled using stochastic distributions and the problem is solved by a two-level method: the master level using a novel genetic algorithm, the slave level using the point estimate method, incorporating the interior point algorithm.

In order to forecast day-ahead power from PV, Grimaccia et al. [4] proposed a general procedure to set up the main characteristics of the network contains number of neurons, layout, and number of trials using a physical hybrid method (PHANN) provided by forecasted meteorological parameters, historical measurements of power production and estimation of clear sky radiation data to perform the day-ahead PV power forecast. The minimum absolute mean error (normalized or weighted) index has been studied to create the most effective configuration for a feed-forward neural network (FFNN). The Levenberg–Marquardt (LM) algorithm is chosen as the training method, together with slow convergence setting.

Harmony search (HS) meta-heuristic algorithm is proposed by Guo et al. [5] after the optimization problem is specified to discover optimum tilt and azimuth angle to maximize extraterrestrial radiation on a collector in China. The results are compared with a reference group which is obtained by the ergodic method conducted in different cities to understand the performance of HS. Additionally, particle swarm optimization (PSO) is used to compare the solution quality with the HS algorithm.

Petrone et al. [6] proposed a genetic algorithm(GA) to obtain the exact solution for SDM parameter identification requiring only some measured points close to a maximum power point (MPP).

Xiong et al. [7] use a symbiotic organisms search algorithm (SOS) to extract parameters from solar cell models. The effectiveness of this model validated by the single diode model, double diode model, and PV module model. In addition, to verify the effectiveness of SOS, five state-of-the-art algorithms including an across neighborhood search (ANS) biogeography-based learning particle swarm optimization (BLPSO), competitive swarm optimizer (CSO), chaotic teaching-learning algorithm (CTLA), and levy flight trajectory-based whale optimization algorithm (LWOA) are used for performance comparison. Comparison on a statistical level is done by the Wilcoxon's rank sum test at a 0.05 confidence level to identify the significance difference between SOS and other compared methods on the same case.

Dolara et al. [8] simulated different dynamic and partial shading conditions based on the PSO evolutionary approach maximum power point tracking (MPPT) algorithm and compared it with classical MPPT methods to investigate conversion efficiency in the conducted scenarios.

Dolara et al. [9] analyzed different approaches in training data set composition for ANN to be used in the physical hybrid method. For ANN, the training algorithm is chosen as Levenberg–Marquardt while the activation function is sigmoid and the number of trials in the ensemble forecast is 40. An additional performance index (envelope-weighted mean absolute error) is proposed to compare results between different approaches.

Mohamed Louzazni et al. [10] perform a comparison among bioinspired algorithms by taking three cases: single diode model, double diode model, and photovoltaic module to predict solar cell and PV module parameters. The Firefly algorithm is chosen for the optimization problem. The results are compared with recent techniques such as the biogeography-based optimization algorithm with mutation strategies (BBO-M), the Levenberg–Marquardt algorithm combined with simulated annealing (LMSA), artificial bee swarm optimization algorithm, artificial bee colony optimization (ABC), hybrid Nelder–Mead and modified particle swarm optimization (NMMPSO), repaired adaptive differential evolution (RADE), chaotic asexual reproduction optimization (CARO) for solar cell single and double diodes; quasi-Newton (Q-N) method and self-organizing migrating algorithm (SOMA) for a-Si:H solar cell and the optimal parameters of Photowatt-PWP 201 are compared with the Newton–Raphson pattern search(PS), genetic algorithm (GA) and simulated annealing algorithm(SA). In conclusion, this Special Issue contains a series of up-to-date research work covering a wide area of application-oriented computational intelligence. This collection of ten papers is highly recommended and will benefit readers in various aspects dealing with solar power systems.

Acknowledgments: We would like to thank all authors, the many dedicated referees, the editor team of *Applied Sciences*, and especially Xiaoyan Chen (Assistant Managing Editor), for their valuable contributions, making this special issue a success.

Conflicts of Interest: The authors declare no conflict of interest.

References

1. Jeong, K.; Hong, T.; Koo, C.; Oh, J.; Lee, M.; Kim, J. A Prototype Design and Development of the Smart Photovoltaic System Blind Considering the Photovoltaic Panel, Tracking System, and Monitoring System. *Appl. Sci.* **2017**, *7*, 1077. [CrossRef]
2. Hammami, M.; Torretti, S.; Grimaccia, F.; Grandi, G. Thermal and Performance Analysis of a Photovoltaic Module with an Integrated Energy Storage System. *Appl. Sci.* **2017**, *7*, 1107. [CrossRef]

3. Hong, Y.-Y.; Yo, P.-S. Novel Genetic Algorithm-Based Energy Management in a Factory Power System Considering Uncertain Photovoltaic Energies. *Appl. Sci.* **2017**, *7*, 438. [CrossRef]
4. Grimaccia, F.; Leva, S.; Mussetta, M.; Ogliari, E. ANN Sizing Procedure for the Day-Ahead Output Power Forecast of a PV Plant. *Appl. Sci.* **2017**, *7*, 622. [CrossRef]
5. Guo, M.; Zang, H.; Gao, S.; Chen, T.; Xiao, J.; Cheng, L.; Wei, Z.; Sun, G. Optimal Tilt Angle and Orientation of Photovoltaic Modules Using HS Algorithm in Different Climates of China. *Appl. Sci.* **2017**, *7*, 1028. [CrossRef]
6. Petrone, G.; Luna, M.; la Tona, G.; di Piazza, M.; Spagnuolo, G. Online Identification of Photovoltaic Source Parameters by Using a Genetic Algorithm. *Appl. Sci.* **2018**, *8*, 9. [CrossRef]
7. Xiong, G.; Zhang, J.; Yuan, X.; Shi, D.; He, Y. Application of Symbiotic Organisms Search Algorithm for Parameter Extraction of Solar Cell Models. *Appl. Sci.* **2018**, *8*, 2155. [CrossRef]
8. Dolara, A.; Grimaccia, F.; Mussetta, M.; Ogliari, E.; Leva, S. An evolutionary-based MPPT algorithm for photovoltaic systems under dynamic partial shading. *Appl. Sci.* **2018**, *8*, 558. [CrossRef]
9. Dolara, A.; Grimaccia, F.; Leva, S.; Mussetta, M.; Ogliari, E. Comparison of Training Approaches for Photovoltaic Forecasts by Means of Machine Learning. *Appl. Sci.* **2018**, *8*, 228. [CrossRef]
10. Louzazni, M.; Khouya, A.; Amechnoue, K.; Gandelli, A.; Mussetta, M.; Crăciunescu, A. Metaheuristic Algorithm for Photovoltaic Parameters: Comparative Study and Prediction with a Firefly Algorithm. *Appl. Sci.* **2018**, *8*, 339. [CrossRef]

© 2019 by the authors. Licensee MDPI, Basel, Switzerland. This article is an open access article distributed under the terms and conditions of the Creative Commons Attribution (CC BY) license (http://creativecommons.org/licenses/by/4.0/).

Article

Application of Symbiotic Organisms Search Algorithm for Parameter Extraction of Solar Cell Models

Guojiang Xiong [1,*], Jing Zhang [1], Xufeng Yuan [1], Dongyuan Shi [2] and Yu He [1]

[1] Guizhou Key Laboratory of Intelligent Technology in Power System, College of Electrical Engineering, Guizhou University, Guiyang 550025, China; jingzhanggzu@126.com (J.Z.); xfyuan@gzu.edu.cn (X.Y.); yhe7@gzu.edu.cn (Y.H.)
[2] State Key Laboratory of Advanced Electromagnetic Engineering and Technology, Huazhong University of Science and Technology, Wuhan 430074, China; dongyuanshi401@163.com
* Correspondence: gjxiongee@foxmail.com; Tel.: +86-0851-8362-6560

Received: 23 September 2018; Accepted: 1 November 2018; Published: 4 November 2018

Abstract: Extracting accurate values for relevant unknown parameters of solar cell models is vital and necessary for performance analysis of a photovoltaic (PV) system. This paper presents an effective application of a young, yet efficient metaheuristic, named the symbiotic organisms search (SOS) algorithm, for the parameter extraction of solar cell models. SOS, inspired by the symbiotic interaction ways employed by organisms to improve their overall competitiveness in the ecosystem, possesses some noticeable merits such as being free from tuning algorithm-specific parameters, good equilibrium between exploration and exploitation, and being easy to implement. Three test cases including the single diode model, double diode model, and PV module model are served to validate the effectiveness of SOS. On one hand, the performance of SOS is evaluated by five state-of-the-art algorithms. On the other hand, it is also compared with some well-designed parameter extraction methods. Experimental results in terms of the final solution quality, convergence rate, robustness, and statistics fully indicate that SOS is very effective and competitive.

Keywords: solar photovoltaic; parameter extraction; symbiotic organisms search; metaheuristic

---

1. Introduction

Solar energy is considered as a promising tool to fight environmental pollution and fossil energy consumption. As the main application of solar energy, solar photovoltaic (PV) has recently achieved leapfrog development. Solarpower Europe reveals that only seven countries installed over 1 GW PV in 2016. That number was changed to nine in 2017, and in 2018, the number keeps increasing and should reach 14 [1]. China, as the country with the biggest capacity of PV power, installed 24.3 GW, which was about 38% of the world's newly installed capacity PV power, in the first half of 2018 [2]. According to data from the International Energy Agency, by 2040, the fast-developing market of PV in China and India will cause solar to be the largest source of low-carbon capacity [3]. A PV system is a multi-component power unit utilized to directly convert solar energy into electricity. As the core device of a PV system, a solar cell's accurate modelling and parameter extraction are very important for the performance analysis of the PV system [4]. For solar cells, their current-voltage (*I-V*) characteristics are widely simulated by the most popular single diode model and double diode model [5], which have five and seven unknown parameters, respectively, that need to be extracted.

Extracting accurate value for these relevant unknown model parameters is vital and necessary, and has drawn researchers' attention in recent years [6,7]. The propounded parameter extraction methods roughly include analytical methods [8–15] and optimization methods. Analytical methods employ mathematical formulations to obtain the model parameters based on a few pivotal data points of *I-V* characteristic curve. Their merits are simplicity, computational efficiency, and ease of implementation. However, the solution quality depends heavily on the accuracy of the opted data points. A small degree of noise on these points may result in significant errors for these parameters.

Instead of relying on several key data points, optimization methods take all measured data points into account. The parameter extraction problem is firstly converted to an optimization problem. A well-designed optimization method is then used to solve the problem to optimality with the goal of fitting all measured points. Compared with the analytical methods, the dominant advantage of optimization methods is that more accurate values for these relevant parameters can be achieved as a result of the utilization of all measured *I-V* points. The optimization methods consist of deterministic methods and metaheuristic methods. Deterministic methods, in general, are local search algorithms because they rely mostly on the gradient information. Therefore, they are prone to being caught in a local extremum, especially in solving intricate multimodal problems such as the parameter extraction problem concerned here. In addition, they require the target functions to be convex and differentiable, among others. To meet the implementation demand, simplification and linearization are usually needed, which may lead to poor approximate solutions and thus cause them to be unreliable [16].

Metaheuristic methods, as a feasible and effective alternative to the deterministic methods, have gained increasing interest recently. They relax the problem formulation and pay no attention to the gradient information, and thus can overcome the shortcomings of deterministic methods. Hence, they can serve as reliable tools for multimodal problems. In the last few years, researchers have attempted to apply various metaheuristic methods to deal with the problem concerned in this paper. Bastidas-Rodriguez et al. [17] utilized genetic algorithm (GA) to extract parameters of the single diode model based on five operating points. El-Naggar et al. [18] applied simulated annealing (SA) to identify parameters of PV models. Bana and Saini [19] developed a particle swarm optimization (PSO) with binary constraints to extract single diode model parameters. Nunes et al. [20] proposed a guaranteed convergence PSO for both benchmark cases and real experimental data. Ishaque et al. [21] put forward a penalty based differential evolution (DE) to achieve accurate parameters of PV modules at different environmental conditions. Chellaswamy and Ramesh [22] designed an adaptive DE to yield accurate parameters of solar cell models. Jiang et al. [23] implemented an improved adaptive DE (IADE) to estimate the parameters of solar cells and modules. Askarzadeh and Rezazadeh [24] applied artificial bee swarm optimization (ABSO) to obtain promising parameters for both single diode and double diode models. Chen et al. [25] proposed a generalized oppositional teaching-learning-based optimization (GOTLBO) to acquire accurate parameters of solar cells, and then hybridized artificial bee colony (ABC) with TLBO to identify parameters of different PV models [26]. Yu et al. developed several well-designed methods including self-adaptive TLBO [27], improved JAYA (IJAYA) [28], and multiple learning backtracking search algorithm [29] to estimate parameters of PV models. Oliva et al. used chaotic maps to enhance the performance of whale optimization algorithm (WOA) [30] and ABC [31], respectively, for parameter extraction of solar cells. Kichou et al. [32] employed five different algorithms to achieve parameters for two PV models. Ma et al. [33] statistically compared the performance of six algorithms on parameter extraction of PV models. In addition to the abovementioned methods, many more different types of metaheuristics [34–46] are also applied to the problem considered here.

Metaheuristic methods exhibit diverse attributes regarding number of tuning parameters and searching strategies. However, the famous no-free-lunch theorem [47] has highly remarked that no single method that can be adopted as the gold standard for every optimization problem. Hence, it is necessary and important to attempt new ones with the constant hope of obtaining promising solutions for the parameter extraction problem of solar cell models, which motivates the authors to apply a young, yet efficient metaheuristic named the symbiotic organisms search (SOS) algorithm in this paper to assess its performance. SOS, proposed by Cheng and Prayogo [48], is inspired by the symbiotic interaction ways employed by organisms to improve their overall competitiveness in the ecosystem. SOS has some noticeable merits such as being free from tuning algorithm-specific parameters, good equilibrium between exploration and exploitation, and being easy to implement [49,50]. These merits encourage researchers to apply SOS to a host of engineering problems.

SOS has proven itself a worthy competitor and alternative in many optimization problems. Nonetheless, the promising method has not been employed to solve the problem considered here. The aim of this paper is first to present experimental results validating the performance of SOS in dealing with the parameter extraction problem of solar cell models. Three test cases consisting of the single diode model, double diode model, and PV module model are served to evaluate the effectiveness of SOS along with necessary comparisons. The experimental results comprehensively indicate that SOS behaves competitively compared with other methods.

The rest of this paper is organized as follows. The problem formulation is briefly presented in Section 2. In Section 3, the SOS is provided. Then, the results are analyzed in Section 4 and this paper is concluded in Section 5.

2. Problem Formulation

2.1. Single Diode Model

Single diode model is a very popular model used to simulate the I-V characteristic of a solar cell. The output current I_L (A), as depicted in Figure 1, can be formulated as follows according to Kirchhoff's current law.

$$I_L = I_{ph} - I_d - I_{sh} \tag{1}$$

where I_{ph}, I_d, and I_{sh} are the photo generated current (A), diode current (A), and shunt resistor current (A), respectively. I_d and I_{sh} are calculated by Equations (2) and (3), respectively [24,35,51–53].

$$I_d = I_{sd} \cdot [\exp(\frac{V_L + R_s \cdot I_L}{nV_t}) - 1] \tag{2}$$

$$I_{sh} = \frac{V_L + R_s \cdot I_L}{R_{sh}} \tag{3}$$

$$V_t = \frac{kT}{q} \tag{4}$$

where V_L and V_t represent the output voltage (V) and thermal voltage (V), respectively. I_{sd} is the reverse saturation current (A). R_s and R_{sh} denote the series resistance (Ω) and shunt resistance (Ω), respectively. n is the diode ideal factor. $k = 1.3806503 \times 10^{-23}$ J/K is the Boltzmann constant. $q = 1.60217646 \times 10^{-19}$ C is the electron charge. T denotes the cell temperature in Kelvin.

Substituting Equations (2)–(4) into Equation (1), the output current I_L can be written as follows:

$$I_L = I_{ph} - I_{sd} \cdot [\exp(\frac{V_L + R_s \cdot I_L}{nV_t}) - 1] - \frac{V_L + R_s \cdot I_L}{R_{sh}} \tag{5}$$

It is observed from Equation (5) that if we know the values of I_{ph}, I_{sd}, R_s, R_{sh}, and n, then the I-V characteristic of this model can be constructed. Therefore, accurate extraction of these five unknown parameters is the core of this study.

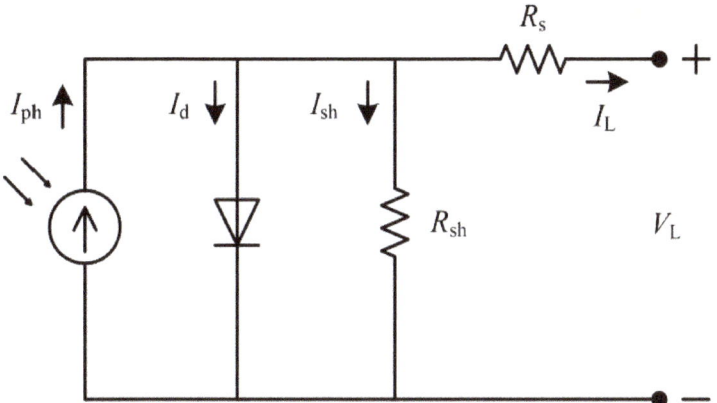

Figure 1. Single diode model.

2.2. Double Diode Model

The above model performs well for almost all types of solar cells [5]. However, its performance is unsatisfactory at low irradiance for thin films based solar cells. The problem can be handled well by the double diode model [24,35]. The output current in Figure 2 is formulated as follows [52,54,55]:

$$\begin{aligned} I_L &= I_{ph} - I_{d1} - I_{d2} - I_{sh} \\ &= I_{ph} - I_{sd1} \cdot [\exp(\frac{V_L + R_s \cdot I_L}{n_1 V_t}) - 1] \\ &\quad - I_{sd2} \cdot [\exp(\frac{V_L + R_s \cdot I_L}{n_2 V_t}) - 1] - \frac{V_L + R_s \cdot I_L}{R_{sh}} \end{aligned} \quad (6)$$

where I_{sd1} and I_{sd2} represent the diffusion current (A) and saturation current (A), respectively. n_1 and n_2 are the diode ideal factors. Compared with the single diode mode, this model adds two more unknown parameters (I_{sd2} and n_2) and thereby the total number of unknown parameters that need to be extracted is seven ($I_{ph}, I_{sd1}, I_{sd2}, R_s, R_{sh}, n_1$ and n_2).

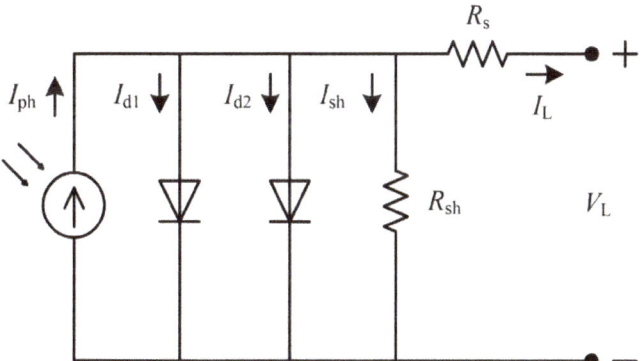

Figure 2. Double diode model.

2.3. PV Module

In general, a PV module is used to raise the output voltage. The corresponding output current is calculated as follows [19,28,56,57]:

$$I_L = N_p \left\{ I_{ph} - I_{sd} \cdot [\exp(\frac{V_L/N_s + R_s I_L N_p}{nV_t}) - 1] - \frac{V_L/N_s + R_s I_L/N_p}{R_{sh}} \right\} \quad (7)$$

where N_s and N_p denote the number of solar cells in series and in parallel, respectively.

2.4. Objective Function

Accurate extracted values for the involved unknown parameters of solar cell models should make the constructed model coincide with the real model. Namely, by using the constructed model, the calculated data should match the measured data well. Therefore, the difference between the measured current and the calculated current can be used to reflect the agreement degree. In general, the root mean square error (RMSE) is highly preferred [18,20–25].

$$\min F(x) = \text{RMSE}(x) = \sqrt{\frac{1}{N}\sum_{i=1}^{N} f_i(V_L, I_L, x)^2} \qquad (8)$$

where N is the number of measured data and x is the solution vector.

For the abovementioned three models, the objective functions $f(V_L, I_L, x)$ and the solution vectors x are as follows:

$$\begin{cases} f_{\text{single diode}}(V_L, I_L, x) = I_{ph} - I_{sd} \cdot [\exp(\frac{V_L+R_s \cdot I_L}{nV_t}) - 1] - \frac{V_L+R_s \cdot I_L}{R_{sh}} - I_L \\ x_{\text{single diode}} = \{I_{ph}, I_{sd}, R_s, R_{sh}, n\} \end{cases} \qquad (9)$$

$$\begin{cases} f_{\text{double diode}}(V_L, I_L, x) = I_{ph} - I_{sd1} \cdot [\exp(\frac{V_L+R_s \cdot I_L}{n_1 V_t}) - 1] \\ \qquad - I_{sd2} \cdot [\exp(\frac{V_L+R_s \cdot I_L}{n_2 V_t}) - 1] \\ \qquad - \frac{V_L+R_s \cdot I_L}{R_{sh}} - I_L \\ x_{\text{double diode}} = \{I_{ph}, I_{sd1}, I_{sd2}, R_s, R_{sh}, n_1, n_2\} \end{cases} \qquad (10)$$

$$\begin{cases} f_{\text{PV module}}(V_L, I_L, x) = N_p \left\{ I_{ph} - I_{sd} \cdot [\exp(\frac{V_L/N_s+R_s I_L/N_p}{nV_t}) - 1] - \frac{V_L/N_s+R_s I_L/N_p}{R_{sh}} \right\} - I_L \\ x_{\text{PV module}} = \{I_{ph}, I_{sd}, R_s, R_{sh}, n\} \end{cases} \qquad (11)$$

3. Symbiotic Organisms Search (SOS) Algorithm

SOS [48] is a young, yet effective metaheuristic inspired by the symbiotic interaction ways employed by organisms to improve their overall competitiveness in the ecosystem. Each organism (i.e., population individual) is represented as a D-dimensional vector $X_i = [x_{i,1}, x_{i,2}, \ldots, x_{i,D}]$, where $i = 1, 2, \ldots, ps$, ps is the number of organisms in the ecosystem (i.e., population size). SOS contains mutualism, commensalism, and parasitism phases.

3.1. Mutualism Phase

In this phase, two organisms establish a good interaction relationship in which they can obtain what they need, and thus their mutual survival advantage can be increased simultaneously. For each organism X_i of the ecosystem, a random distinct organism X_j is selected to interact with X_i by the following formulations:

$$X_{i,\text{new}} = X_i + \text{rand}(0,1) \cdot (X_{\text{best}} - BF_1 \cdot MV) \qquad (12)$$

$$X_{j,\text{new}} = X_j + \text{rand}(0,1) \cdot (X_{\text{best}} - BF_2 \cdot MV) \qquad (13)$$

where $X_{i,\text{new}}$ and $X_{j,\text{new}}$ are new candidate solutions for X_i and X_j, respectively. $\text{rand}(a,b)$ is a random number generated uniformly in (a,b). BF_1 and BF_2 are benefit factors with the random value 1 or 2. X_{best} represents the best organism of the ecosystem. $MV = (X_i + X_j)/2$ is the relationship characteristic.

3.2. Commensalism Phase

In this phase, two organisms build a unidirectional relationship where one organism X_i benefits from the other organism X_j as shown in Equation (14), whereas X_j gets nothing from X_i.

$$X_{i,\text{new}} = X_i + \text{rand}(-1,1) \cdot (X_{\text{best}} - X_j) \tag{14}$$

3.3. Parasitism Phase

In parasitism, one organism X_i improves its survivability through harming the other organism X_j. In SOS, this relationship is modeled as follows. An organism X_i is copied and used to create an artificial parasite AP. Then, some random dimensions of AP are selected and modified by a random number generated within the corresponding bounds. The other organism X_j, selected randomly from the ecosystem, serves as a host to the parasite AP. If AP is better than X_j, then X_j will be replaced by AP; otherwise, AP will be discarded.

The pseudo-code of SOS is presented in Algorithm 1. It can be seen that apart from the common parameter, that is, the population size used in all metaheuristic algorithms, SOS has no algorithm-specific parameters that need to be well-tuned.

Algorithm 1: The pseudo-code of SOS

1:   Initialize an ecosystem X with ps organisms randomly
2:   Calculate the fitness value of each organism
3:   Set the iteration number $t = 1$
4:   **While** the terminating criterion is not met **do**
5:     Select the fittest organism X_{best} of the ecosystem
6:     **For** $i = 1$ to ps **do**
7:       /* mutualism phase */
8:       Select a random organism X_j ($j \neq i$) from the ecosystem
9:       Generate the i-th new organism $X_{i,\text{new}}$ using Equation (12)
10:      Generate the j-th new organism $X_{j,\text{new}}$ using Equation (13)
11:      Calculate the fitness value of $X_{i,\text{new}}$ and $X_{j,\text{new}}$
12:      Replace the old organism if it is defeated by the new one
13:      /* commensalism phase */
14:      Select a random organism X_j ($j \neq i$) from the ecosystem
15:      Generate the i-th new organism $X_{i,\text{new}}$ using Equation (14)
16:      Calculate the fitness value of $X_{i,\text{new}}$
17:      Replace the old organism if it is defeated by the new one
18:      /* parasitism phase */
19:      Select a random organism X_j ($j \neq i$) from the ecosystem
20:      Generate an artificial parasite $AP = X_i$
21:      Select a random number of dimensions of AP
22:      Replace the selected dimensions using a random number
23:      Calculate the fitness value of the modified AP
24:      Replace X_j if the modified AP is better than X_j
25:     End for
26:     $t = t + 1$
27:   End while

---

4. Results and Discussions

4.1. Test PV Models

In this work, SOS is applied to three cases including single diode, double diode, and PV module models. The datasets are derived from the literature [58]. The measurements are conducted on an

RTC France silicon solar cell and a Photowatt-PWP201 solar module. The former operates under 1000 W/m² at 33 °C. The latter contains 36 polycrystalline silicon cells connected in series operating under 1000 W/m² at 45 °C. The boundaries of extracted parameters are presented in Table 1.

Table 1. Parameter boundaries of solar cell models.

Parameter	Single/Double Diode Model		PV Module Model	
	Lower Bound	Upper Bound	Lower Bound	Upper Bound
I_{ph} (A)	0	1	0	2
I_{sd} (μA)	0	1	0	50
R_s (Ω)	0	0.5	0	2
R_{sh} (Ω)	0	100	0	2000
n, n_1, n_2	1	2	1	50

4.2. Experimental Settings

In this work, the maximum number of fitness evaluations (Max_FEs), which is set to 50,000 [29], serves as the terminating criterion. In addition, to verify the effectiveness of SOS, five state-of-the-art algorithms including across neighborhood search (ANS) [59], biogeography-based learning particle swarm optimization (BLPSO) [60], competitive swarm optimizer (CSO) [61], chaotic teaching-learning algorithm (CTLA) [62], and levy flight trajectory-based whale optimization algorithm (LWOA) [63] are used for performance comparison. These five methods keep the original algorithm parameters, except the population size *ps*, setting the same unified value 50 for fair comparison. For each case, each method runs 50 times independently.

4.3. Experimental Results and Comparison

4.3.1. Results Comparison on the Single Diode Model

The experimental results of the first case are tabulated in Table 2. The symbols Min, Max, Mean, and Std. dev. represent the minimum, maximum, mean, and standard deviation values, respectively, over 50 independent runs. The experimental results of some well-designed methods, including SA [18] IADE [23], ABSO [24], GOTLBO [25], IJAYA [28], differential evolution (DE) [33], biogeography-based optimization algorithm with mutation strategies (BBO-M) [34], grouping-based global harmony search (GGHS) [35], chaotic asexual reproduction optimization (CARO) [40], bird mating optimizer (BMO) [44], and pattern search (PS) [45], are also provided in Table 2 for comparison. It can be seen that, compared with ANS, BLPSO, CSO, CTLA, and LWOA, SOS can acquire the lowest RMSE value (9.8609×10^{-4}). Considering the mean, maximum, and standard deviation values, SOS also consistently performs better than them. In addition, SOS is also highly competitive against other recently proposed methods. It is better than IADE, ABSO, BBO-M, GGHS, GOTLBO, CARO, PS, and SA, except not better than DE, IJAYA, and BMO. Although DE, IJAYA, and BMO beat SOS, the disparities are very small.

The best extracted values for the five unknown parameters of single diode model are given in Table 3. We observe that these listed methods almost extract close values for the unknown parameters. Utilizing the extracted parameters in Table 3, we reconstruct the characteristic curves as illustrated in Figure 3. We see that both the output current and power calculated by SOS match the measured values well throughout the whole range of voltage. In addition, we also tabulate the output current data calculated by ANS, BLPSO, CSO, CTLA, LWOA, and SOS in Table 4. An error index the sum of individual absolute error (SIAE) given in Equation (15) is used to evaluate the fitting error. It is obvious that the SIAE value of SOS is the smallest, followed by that of ANS, LWOA, BLPSO, CTLA, and CSO, meaning that SOS achieves more accurate values for the relevant parameters of single diode model.

$$\text{SIAE} = \sum_i^N \left| I_{L_i,measured} - I_{L_i,calculated} \right| \tag{15}$$

Table 2. RMSE results for the single diode model.

Method	Min	Max	Mean	Std. Dev.
IADE	9.8900×10^{-4}	NA	NA	NA
ABSO	9.9124×10^{-4}	NA	NA	NA
BBO-M	9.8634×10^{-4}	NA	NA	NA
GGHS	9.9078×10^{-4}	NA	NA	NA
GOTLBO	9.87442×10^{-4}	1.98244×10^{-3}	1.33488×10^{-3}	2.99407×10^{-4}
CARO	9.8665×10^{-4}	NA	NA	NA
DE	9.8602×10^{-4}	NA	NA	NA
IJAYA	9.8603×10^{-4}	1.0622×10^{-3}	9.9204×10^{-4}	1.4033×10^{-5}
BMO	9.8608×10^{-4}	NA	NA	NA
PS	2.863×10^{-1}	NA	NA	NA
SA	1.70×10^{-3}	NA	NA	NA
ANS	9.9689×10^{-4}	1.4385×10^{-3}	1.1051×10^{-3}	1.0141×10^{-4}
BLPSO	1.4836×10^{-3}	2.2415×10^{-3}	1.9092×10^{-3}	1.7404×10^{-4}
CSO	1.6358×10^{-3}	2.4104×10^{-3}	2.0058×10^{-3}	1.7398×10^{-4}
CTLA	1.0991×10^{-3}	1.8027×10^{-3}	1.3772×10^{-3}	1.7132×10^{-4}
LWOA	1.0873×10^{-3}	9.1622×10^{-3}	3.1119×10^{-3}	1.8838×10^{-3}
SOS	9.8609×10^{-4}	1.1982×10^{-3}	1.0245×10^{-3}	5.2184×10^{-5}

NA: Not available in the literature.

Table 3. xtracted parameters for the single diode model.

Method	I_{ph} (A)	I_{sd} (μA)	R_s (Ω)	R_{sh} (Ω)	n	RMSE
IADE	0.7607	0.33613	0.03621	54.7643	1.4852	9.8900×10^{-4}
ABSO	0.76080	0.30623	0.03659	52.2903	1.47583	9.9124×10^{-4}
BBO-M	0.76078	0.31874	0.03642	53.36277	1.47984	9.8634×10^{-4}
GGHS	0.76092	0.32620	0.03631	53.0647	1.48217	9.9079×10^{-4}
GOTLBO	0.760780	0.331552	0.036265	54.115426	1.483820	9.8744×10^{-4}
CARO	0.76079	0.31724	0.03644	53.0893	1.48168	9.8665×10^{-4}
DE	0.7608	0.323	0.0364	53.719	1.4812	9.8602×10^{-4}
IJAYA	0.7608	0.3228	0.0364	53.7595	1.4811	9.8603×10^{-4}
PS	0.7617	0.9980	0.0313	64.1026	1.6000	2.863×10^{-1}
SA	0.7620	0.4798	0.0345	43.1034	1.5172	1.70×10^{-3}
ANS	0.7607	0.3407	0.0362	54.7917	1.4866	9.9689×10^{-4}
BLPSO	0.7599	0.4977	0.0347	96.5115	1.5257	1.4836×10^{-3}
CSO	1.0205	0.3658	1.2122	1689.0050	48.8206	1.6358×10^{-3}
CTLA	0.7650	0.4280	0.0357	61.1131	1.5092	1.0991×10^{-3}
LWOA	1.0284	0.3145	1.2218	1272.0197	48.2413	1.0873×10^{-3}
SOS	0.7608	0.3579	0.0359	53.7835	1.4916	9.8609×10^{-4}

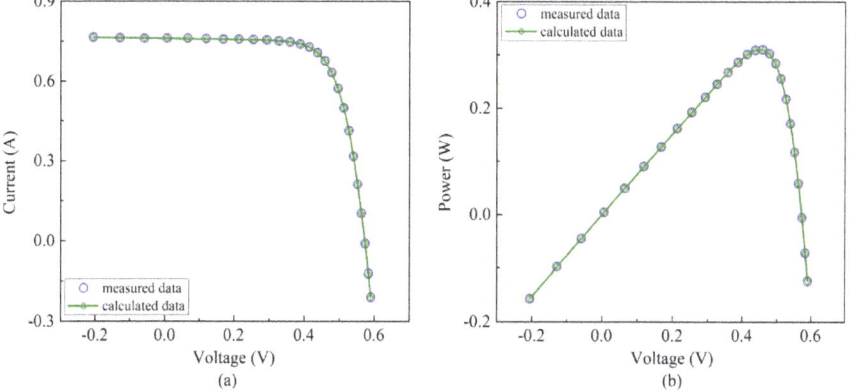

Figure 3. Extraction results by SOS for the single diode model. (a) Current; (b) power.

Table 4. Fitting results for the single diode model.

Item	V_L (V)	I_L Measured (A)	I_L Calculated (A)					
			ANS	BLPSO	CSO	CTLA	LWOA	SOS
1	−0.2057	0.7640	0.7639	0.7617	0.7614	0.7679	0.7631	0.7641
2	−0.1291	0.7620	0.7625	0.7609	0.7606	0.7667	0.7618	0.7627
3	−0.0588	0.7605	0.7613	0.7602	0.7598	0.7655	0.7607	0.7614
4	0.0057	0.7605	0.7601	0.7595	0.7591	0.7645	0.7597	0.7602
5	0.0646	0.7600	0.7590	0.7589	0.7585	0.7635	0.7587	0.7591
6	0.1185	0.7590	0.7580	0.7584	0.7579	0.7626	0.7578	0.7581
7	0.1678	0.7570	0.7571	0.7578	0.7573	0.7618	0.7570	0.7572
8	0.2132	0.7570	0.7561	0.7572	0.7567	0.7609	0.7561	0.7562
9	0.2545	0.7555	0.7551	0.7564	0.7559	0.7599	0.7552	0.7551
10	0.2924	0.7540	0.7537	0.7552	0.7546	0.7585	0.7538	0.7537
11	0.3269	0.7505	0.7514	0.7530	0.7524	0.7561	0.7517	0.7514
12	0.3585	0.7465	0.7473	0.7489	0.7484	0.7519	0.7477	0.7473
13	0.3873	0.7385	0.7400	0.7414	0.7412	0.7443	0.7406	0.7399
14	0.4137	0.7280	0.7273	0.7282	0.7288	0.7310	0.7280	0.7271
15	0.4373	0.7065	0.7068	0.7072	0.7091	0.7099	0.7076	0.7065
16	0.4590	0.6755	0.6751	0.6750	0.6789	0.6774	0.6759	0.6748
17	0.4784	0.6320	0.6306	0.6303	0.6369	0.6322	0.6315	0.6303
18	0.4960	0.5730	0.5719	0.5715	0.5812	0.5727	0.5726	0.5716
19	0.5119	0.4990	0.4994	0.4991	0.5119	0.4997	0.4999	0.4991
20	0.5265	0.4130	0.4134	0.4134	0.4288	0.4133	0.4137	0.4133
21	0.5398	0.3165	0.3173	0.3175	0.3342	0.3169	0.3173	0.3172
22	0.5521	0.2120	0.2122	0.2126	0.2292	0.2116	0.2120	0.2122
23	0.5633	0.1035	0.1029	0.1032	0.1181	0.1021	0.1026	0.1029
24	0.5736	−0.0100	−0.0091	−0.0091	−0.0025	−0.0101	−0.0094	−0.0091
25	0.5833	−0.1230	−0.1243	−0.1249	−0.1180	−0.1255	−0.1245	−0.1244
26	0.5900	−0.2100	−0.2092	−0.2104	−0.2078	−0.2105	−0.2092	−0.2094
	SIAE		0.0182	0.0275	0.1347	0.0739	0.0191	0.0181

Besides, the convergence curves are presented in Figure 4. It is obvious that SOS is slightly slower than LWOA in the opening phase, however, the latter stagnates soon and then suffers from premature convergence, indicating that it has been caught in a local optimum. For the other four methods, SOS consistently converges faster than them throughout the whole evolutionary process.

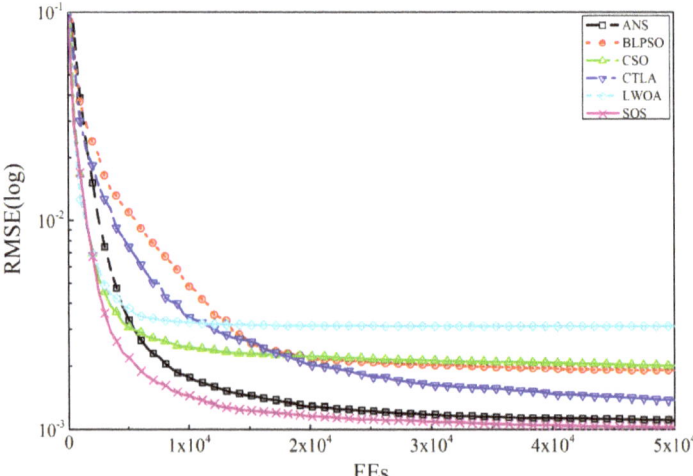

Figure 4. Convergence curves for the single diode model.

4.3.2. Results Comparison on the Double Diode Model

The experimental results of the second case are summarized in Table 5. Similar to the comparison results on the single diode model, SOS performs better than ANS, BLPSO, CSO, CTLA, and LWOA in various RMSE indicators on the double diode model. SOS is surpassed by GOTLBO, CARO, and IJAYA, but it outperforms GGHS, PS, and SA. It is worth noting that the standard deviation value of SOS is the smallest among all compared methods, which indicates that SOS is highly robust. The extracted parameters are tabulated in Table 6. The reconstructed characteristic curves provided in Figure 5 clearly demonstrate that the calculated current and power achieved by SOS match up well with the measured values. The curve fitting results presented in Table 7 manifest once again that SOS can yield the smallest SIAE value (0.0182), followed by ANS, LWOA, BLPSO, CTLA, and CSO, which demonstrates the high accuracy of the parameters extracted by SOS for the double diode model. The convergence graph illustrated in Figure 6 reveals that SOS exhibits noticeably faster convergence rate than BLPSO, CSO, CTLA, and LWOA, but not ANS, which is slightly faster than SOS during the intermediate stage. However, ANS is surpassed by SOS in other stages.

Table 5. RMSE results for the double diode model.

Method	Min	Max	Mean	Std. dev.
GGHS	9.8635×10^{-4}	NA	NA	NA
GOTLBO	9.83177×10^{-4}	1.78774×10^{-3}	1.24360×10^{-3}	2.09115×10^{-4}
CARO	9.8260×10^{-4}	NA	NA	NA
IJAYA	9.8293×10^{-4}	1.4055×10^{-3}	1.0269×10^{-3}	9.8625×10^{-5}
PS	1.5180×10^{-2}	NA	NA	NA
SA	1.9000×10^{-2}	NA	NA	NA
ANS	1.0042×10^{-3}	1.4456×10^{-3}	1.1337×10^{-3}	9.9500×10^{-5}
BLPSO	1.5704×10^{-3}	2.5312×10^{-3}	2.0554×10^{-3}	2.0186×10^{-4}
CSO	1.7013×10^{-3}	2.7735×10^{-3}	2.2421×10^{-3}	2.2059×10^{-4}
CTLA	1.3216×10^{-3}	3.1002×10^{-3}	2.0145×10^{-3}	4.0895×10^{-4}
LWOA	1.3120×10^{-3}	1.3387×10^{-2}	3.5838×10^{-3}	2.6270×10^{-3}
SOS	9.8518×10^{-4}	1.3498×10^{-3}	1.0627×10^{-3}	9.6141×10^{-5}

NA: Not available in the literature.

Table 6. Extracted parameters for the double diode model.

Method	I_{ph} (A)	I_{sd1} (μA)	R_s (Ω)	R_{sh} (Ω)	n_1	I_{sd2} (μA)	n_2	RMSE
GGHS	0.76079	0.97310	0.03690	56.8368	1.92126	0.16791	1.42814	9.8635×10^{-4}
GOTLBO	0.760752	0.800195	0.036783	56.075304	1.999973	0.220462	1.448974	9.83177×10^{-4}
CARO	0.76075	0.29315	0.03641	54.3967	1.47338	0.09098	1.77321	9.8260×10^{-4}
IJAYA	0.7601	0.0050445	0.0376	77.8519	1.2186	0.75094	1.6247	9.8293×10^{-4}
PS	0.7602	0.9889	0.0320	81.3008	1.6000	0.0001	1.1920	1.5180×10^{-2}
SA	0.7623	0.4767	0.0345	43.1034	1.5172	0.0100	2.0000	1.9000×10^{-2}
ANS	0.7609	0.1785	0.0369	51.5905	1.8181	0.2466	1.4581	1.0042×10^{-3}
BLPSO	0.7607	0.5481	0.0338	78.6922	1.5442	0.0542	1.5765	1.5704×10^{-3}
CSO	0.7628	0.7954	0.0409	15.7733	1.6936	0.6780	1.8138	1.7013×10^{-3}
CTLA	0.7570	0.8542	0.0313	89.6464	1.7879	0.3812	1.5230	1.3216×10^{-3}
LWOA	0.7597	0.2342	0.0355	86.8763	1.4679	0.3709	1.6989	1.3120×10^{-3}
SOS	0.7606	0.5408	0.0365	55.5537	1.9346	0.2418	1.4579	9.8518×10^{-4}

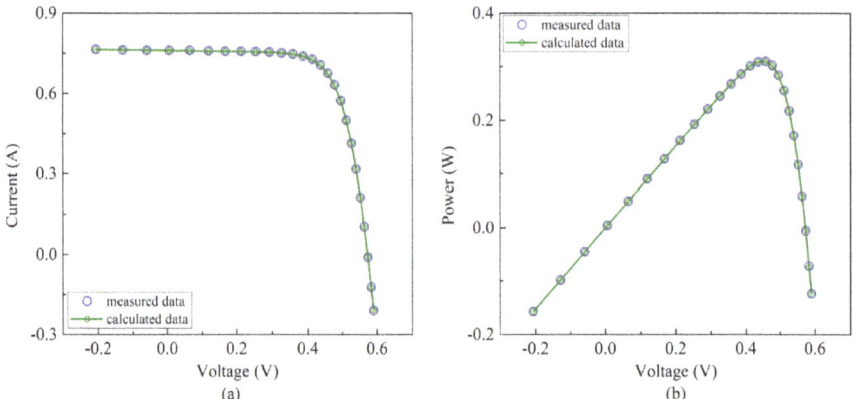

Figure 5. Extraction results by SOS for the double diode model. (a) Current; (b) power.

Table 7. Fitting results for the double diode model.

Item	V_L (V)	I_L Measured (A)	I_L Calculated (A)					
			ANS	BLPSO	CSO	CTLA	LWOA	SOS
1	−0.2057	0.7640	0.7644	0.7630	0.7738	0.7591	0.7618	0.7638
2	−0.1291	0.7620	0.7629	0.7620	0.7690	0.7582	0.7609	0.7625
3	−0.0588	0.7605	0.7615	0.7611	0.7645	0.7574	0.7601	0.7612
4	0.0057	0.7605	0.7603	0.7603	0.7605	0.7567	0.7593	0.7600
5	0.0646	0.7600	0.7591	0.7595	0.7567	0.7561	0.7586	0.7590
6	0.1185	0.7590	0.7581	0.7588	0.7533	0.7554	0.7580	0.7580
7	0.1678	0.7570	0.7571	0.7581	0.7501	0.7548	0.7574	0.7571
8	0.2132	0.7570	0.7561	0.7574	0.7470	0.7541	0.7567	0.7561
9	0.2545	0.7555	0.7550	0.7565	0.7440	0.7532	0.7559	0.7551
10	0.2924	0.7540	0.7536	0.7552	0.7407	0.7518	0.7547	0.7536
11	0.3269	0.7505	0.7513	0.7528	0.7366	0.7494	0.7525	0.7513
12	0.3585	0.7465	0.7472	0.7485	0.7312	0.7449	0.7484	0.7472
13	0.3873	0.7385	0.7400	0.7407	0.7233	0.7369	0.7410	0.7399
14	0.4137	0.7280	0.7274	0.7273	0.7116	0.7234	0.7280	0.7271
15	0.4373	0.7065	0.7071	0.7060	0.6951	0.7023	0.7072	0.7066
16	0.4590	0.6755	0.6756	0.6737	0.6718	0.6704	0.6753	0.6750
17	0.4784	0.6320	0.6312	0.6290	0.6412	0.6265	0.6307	0.6307
18	0.4960	0.5730	0.5724	0.5704	0.6024	0.5689	0.5719	0.5720
19	0.5119	0.4990	0.4997	0.4984	0.5557	0.4979	0.4994	0.4995
20	0.5265	0.4130	0.4136	0.4133	0.5009	0.4136	0.4137	0.4136
21	0.5398	0.3165	0.3172	0.3178	0.4394	0.3185	0.3176	0.3174
22	0.5521	0.2120	0.2120	0.2133	0.3717	0.2137	0.2126	0.2123
23	0.5633	0.1035	0.1026	0.1041	0.3002	0.1034	0.1031	0.1029
24	0.5736	−0.0100	−0.0093	−0.0082	0.2259	−0.0105	−0.0090	−0.0091
25	0.5833	−0.1230	−0.1243	−0.1241	0.1483	−0.1289	−0.1246	−0.1243
26	0.5900	−0.2100	−0.2089	−0.2098	0.0904	−0.2168	−0.2098	−0.2091
	SIAE		0.0189	0.0283	1.6176	0.0789	0.0247	0.0182

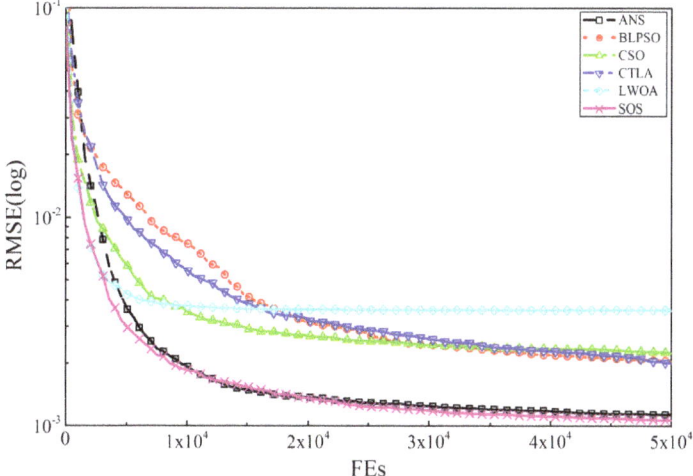

Figure 6. Convergence curves for the double diode model.

4.3.3. Results Comparison on the PV Module Model

The RMSE values of the third case listed in Table 8 indicate that SOS, together with IJAYA, can provide the smallest RMSE value (2.4251×10^{-3}) among all methods. Based on the optimal extracted parameters in Table 9, the corresponding characteristic curves are rebuilt and illustrated in Figure 7. It is clear that the output current and power calculated by SOS are highly in coincidence with the measured values. The SIAE results presented in Table 10 repeatedly manifest that SOS can achieve the most accurate values for the unknown parameters, followed by ANS, BLPSO, LWOA, CTLA, and CSO. The curves presented in Figure 8 state clearly that SOS is consistently faster than its competitors from beginning to end.

Table 8. RMSE results for the photovoltaic (PV) module model.

Method	Min	Max	Mean	Std. dev.
CARO	2.427×10^{-3}	NA	NA	NA
IJAYA	2.4251×10^{-3}	2.4393×10^{-3}	2.4289×10^{-3}	3.7755×10^{-6}
PS	1.18×10^{-2}	NA	NA	NA
SA	2.70×10^{-3}	NA	NA	NA
ANS	2.4310×10^{-3}	2.5658×10^{-3}	2.4702×10^{-3}	2.9121×10^{-5}
BLPSO	2.4296×10^{-3}	2.5616×10^{-3}	2.4884×10^{-3}	3.3055×10^{-5}
CSO	2.4537×10^{-3}	3.0650×10^{-3}	2.5804×10^{-3}	7.7274×10^{-5}
CTLA	2.4782×10^{-3}	3.5579×10^{-3}	2.7760×10^{-3}	2.4714×10^{-4}
LWOA	2.6352×10^{-3}	6.7023×10^{-2}	1.0936×10^{-2}	1.3115×10^{-2}
SOS	2.4251×10^{-3}	2.5103×10^{-3}	2.4361×10^{-3}	1.7503×10^{-5}

NA: Not available in the literature.

Table 9. Extracted parameters for the PV module model.

Method	I_{ph} (A)	I_{sd} (µA)	R_s (Ω)	R_{sh} (Ω)	n	RMSE
CARO	1.03185	3.28401	1.20556	841.3213	48.40363	2.427×10^{-3}
IJAYA	1.0305	3.4703	1.2016	977.3752	48.6298	2.4251×10^{-3}
PS	1.0313	3.1756	1.2053	714.2857	48.2889	1.18×10^{-2}
SA	1.0331	3.6642	1.1989	833.3333	48.8211	2.7000×10^{-3}
ANS	1.0301	3.6650	1.1967	1070.4564	48.8377	2.4310×10^{-3}
BLPSO	1.0302	3.6462	1.1964	1029.5378	48.8198	2.4296×10^{-3}
CSO	1.0205	3.6578	1.2122	1689.0050	48.8206	2.4537×10^{-3}
CTLA	1.0248	2.6365	1.2689	1722.6637	47.5838	2.4782×10^{-3}
LWOA	1.0284	3.1435	1.2218	1272.0197	48.2413	2.6352×10^{-3}
SOS	1.0303	3.5616	1.1991	1017.7000	48.7291	2.4251×10^{-3}

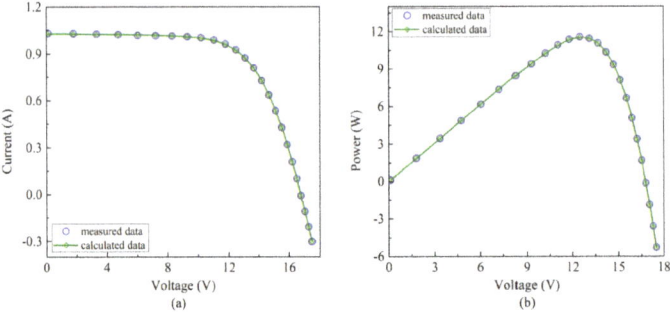

Figure 7. Extraction results by SOS for the photovoltaic (PV) module model. (**a**) Current; (**b**) power.

Table 10. Fitting results for the PV module model.

Item	V_L (V)	I_L Measured (A)	I_L Calculated (A)					
			ANS	BLPSO	CSO	CTLA	LWOA	SOS
1	0.1248	1.0315	1.0288	1.0289	1.0197	1.0240	1.0273	1.0289
2	1.8093	1.0300	1.0272	1.0272	1.0187	1.0230	1.0259	1.0272
3	3.3511	1.0260	1.0257	1.0257	1.0177	1.0220	1.0246	1.0256
4	4.7622	1.0220	1.0241	1.0241	1.0166	1.0210	1.0233	1.0241
5	6.0538	1.0180	1.0224	1.0223	1.0154	1.0198	1.0218	1.0223
6	7.2364	1.0155	1.0201	1.0200	1.0135	1.0180	1.0198	1.0199
7	8.3189	1.0140	1.0166	1.0164	1.0103	1.0151	1.0165	1.0164
8	9.3097	1.0100	1.0108	1.0106	1.0047	1.0098	1.0110	1.0105
9	10.2163	1.0035	1.0009	1.0007	0.9951	1.0006	1.0014	1.0007
10	11.0449	0.9880	0.9848	0.9846	0.9792	0.9850	0.9857	0.9847
11	11.8018	0.9630	0.9598	0.9596	0.9542	0.9603	0.9609	0.9597
12	12.4929	0.9255	0.9230	0.9229	0.9175	0.9235	0.9242	0.9230
13	13.1231	0.8725	0.8725	0.8724	0.8668	0.8726	0.8736	0.8725
14	13.6983	0.8075	0.8072	0.8071	0.8014	0.8064	0.8080	0.8072
15	14.2221	0.7265	0.7278	0.7277	0.7220	0.7261	0.7283	0.7279
16	14.6995	0.6345	0.6363	0.6363	0.6305	0.6337	0.6364	0.6364
17	15.1346	0.5345	0.5356	0.5356	0.5299	0.5323	0.5353	0.5357
18	15.5311	0.4275	0.4288	0.4288	0.4234	0.4252	0.4281	0.4288
19	15.8929	0.3185	0.3186	0.3187	0.3137	0.3154	0.3179	0.3187
20	16.2229	0.2085	0.2079	0.2079	0.2034	0.2053	0.2071	0.2079
21	16.5241	0.1010	0.0984	0.0984	0.0945	0.0970	0.0978	0.0984
22	16.7987	−0.0080	−0.0082	−0.0081	−0.0114	−0.0081	−0.0085	−0.0081
23	17.0499	−0.1110	−0.1110	−0.1110	−0.1135	−0.1093	−0.1109	−0.1109
24	17.2793	−0.2090	−0.2092	−0.2092	−0.2110	−0.2056	−0.2087	−0.2091
25	17.4885	−0.3030	−0.3021	−0.3021	−0.3032	−0.2966	−0.3011	−0.3020
		SIAE	0.0423	0.0424	0.1380	0.0646	0.0452	0.0421

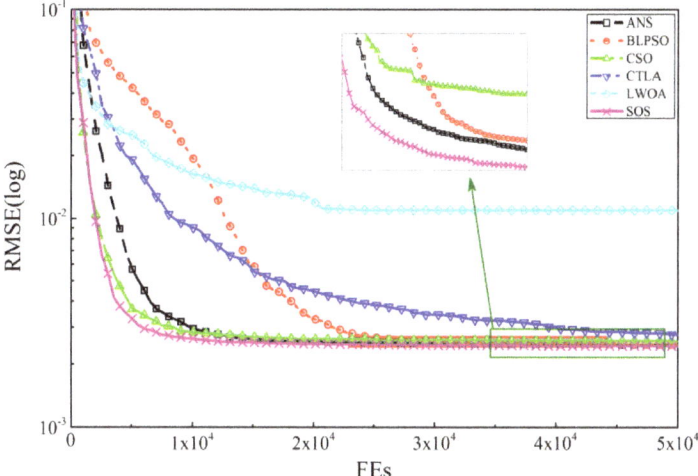

Figure 8. Convergence curves for the PV module model.

4.3.4. Statistical Analysis

The significance difference between two methods can be measured by the statistical analysis. Wilcoxon's rank sum test is a reliable and robust statistical analysis tool and is widely used in metaheuristic methods. In this paper, the Wilcoxon's rank sum test at a 0.05 confidence level is used to identify the significance difference between SOS and other compared methods on the same case. The test results are tabulated in Table 11. The symbol "✝" denotes that SOS is statistically better than its competitor. The results demonstrate that SOS significantly outperforms every method on every case ($p < 0.05$), indicating the better performance of SOS from another perspective.

Table 11. Statistical analysis results based on Wilcoxon's rank sum.

SOS Vs.	Single Diode Model	Double Diode Model	PV Module Model
ANS	✝ ($p = 2.3044 \times 10^{-8}$)	✝ ($p = 3.4341 \times 10^{-6}$)	✝ ($p = 5.5646 \times 10^{-12}$)
BLPSO	✝ ($p = 7.0661 \times 10^{-18}$)	✝ ($p = 7.0661 \times 10^{-18}$)	✝ ($p = 9.9263 \times 10^{-14}$)
CSO	✝ ($p = 7.0661 \times 10^{-18}$)	✝ ($p = 7.0661 \times 10^{-18}$)	✝ ($p = 8.9852 \times 10^{-18}$)
CTLA	✝ ($p = 2.1975 \times 10^{-17}$)	✝ ($p = 7.5041 \times 10^{-18}$)	✝ ($p = 9.5403 \times 10^{-18}$)
LWOA	✝ ($p = 1.2866 \times 10^{-17}$)	✝ ($p = 8.4620 \times 10^{-18}$)	✝ ($p = 7.0661 \times 10^{-18}$)

5. Conclusions and Future Work

The SOS algorithm is applied to solve the parameter extraction problem of solar cell models in this paper. To validate the effectiveness of SOS, it is applied to three models including single diode model, double diode model, and PV module models. From the comparison results of SOS with five state-of-the-art algorithms, namely, ANS, BLPSO, CSO, CTLA, and LWOA, it is summarized that SOS can extract more accurate and robust values for the unknown parameters with a faster convergence rate. The superiority of SOS is also demonstrated through statistical analysis based on the Wilcoxon's rank sum test. In addition, the feasibility of SOS is further confirmed through comparison with some well-designed parameter extraction methods and it indicates that SOS is highly competitive. Meanwhile, there is still room for improvement for SOS to achieve more accurate values, especially for the double diode model. In summary, SOS behaves potential effectively in solving the parameter extraction problem of solar cell models. In future, some advanced strategies such as orthogonal learning and hybridization will be employed to further improve its performance.

Author Contributions: G.X. developed the idea behind the present research and wrote the manuscript. J.Z. provided theoretical analysis. X.Y. and D.S. contributed to final review and manuscript corrections. Y.H. collaborated in processing the experimental data.

Funding: This research was funded by the National Natural Science Foundation of China (Grant No. 51867005, 51667007), the Scientific Research Foundation for the Introduction of Talent of Guizhou University (Grant No. [2017]16), the Guizhou Education Department Growth Foundation for Youth Scientific and Technological Talents (Grant No. QianJiaoHe KY Zi [2018]108), the Guizhou Province Science and Technology Innovation Talent Team Project (Grant No. [2018]5615), the Science and Technology Foundation of Guizhou Province (Grant No. [2016]1036), and the Key Science and Technology Projects of China Southern Power Grid Co., Ltd. (Grant No. GZKJQQ00000417).

Conflicts of Interest: The authors declare no conflict of interest.

Nomenclature

AP	artificial parasite
BF_1, BF_2	benefit factors determined randomly as either 1 or 2
D	dimension of individual vector
I_d	diode current (A)
I_L	output current (A)
I_{ph}	photo generated current (A)
I_{sd}, I_{sd1}, I_{sd2}	saturation currents (A)
I_{sh}	shunt resistor current (A)
k	Boltzmann constant ($1.3806503 \times 10^{-23}$ J/K)
n, n_1, n_2	diode ideality factors
Max_FEs	maximum number of fitness evaluations
N	number of experimental data
N_p	number of cells connected in parallel
N_s	number of cells connected in series
ps	size of population
q	electron charge ($1.60217646 \times 10^{-19}$ C)
rand(a,b)	uniformly distributed random real number in (a,b)
R_s	series resistance (Ω)
R_{sh}	shunt resistance (Ω)
t	current iteration
T	cell temperature (K)
V_L	output voltage (V)
V_t	diode thermal voltage (V)
x	extracted parameters vector
$x_{i,d}$	dth parameter of ith organism
X_i	ith organism
X_{best}	best organism found so far
I-V	current-voltage
P-V	power-voltage
PV	photovoltaic
RMSE	root mean square error
SIAE	sum of individual absolute error
Min	minimum RMSE
Max	maximum RMSE
Mean	mean RMSE
Std Dev	standard deviation
ABSO	artificial bee swarm optimization
ANS	across neighborhood search
BBO-M	biogeography-based optimization algorithm with mutation strategies
BLPSO	biogeography-based learning particle swarm optimization

BMO	bird mating optimizer
CARO	chaotic asexual reproduction optimization
CSO	competitive swarm optimizer
CTLA	chaotic teaching-learning algorithm
DE	differential evolution
GGHS	grouping-based global harmony search
GOTLBO	generalized oppositional teaching learning based optimization
IADE	improved adaptive DE
IJAYA	improved JAYA
LWOA	levy flight trajectory-based whale optimization algorithm
PS	pattern search
SOS	symbiotic organisms search
SA	simulated annealing

References

1. SolarPower Europe. *SolarPower Europe's Global Solar Market Outlook for Solar Power 2018–2022: Solar Growth Ahead*; SolarPower Europe: Brussels, Belgium, 2018.
2. China Energy Net. Available online: http://www.china5e.com (accessed on 5 September 2018).
3. International Energy Agency. *World Energy Outlook 2017*; International Energy Agency: Paris, France, 2017.
4. Youssef, A.; El-Telbany, M.; Zekry, A. The role of artificial intelligence in photo-voltaic systems design and control: A review. *Renew. Sustain. Energy Rev.* 2017, 78, 72–79. [CrossRef]
5. Chin, V.J.; Salam, Z.; Ishaque, K. Cell modelling and model parameters estimation techniques for photovoltaic simulator application: A review. *Appl. Energy* 2015, 154, 500–519. [CrossRef]
6. Jordehi, A.R. Parameter estimation of solar photovoltaic (PV) cells: A review. *Renew. Sustain. Energy Rev.* 2016, 61, 354–371. [CrossRef]
7. Rhouma, M.B.H.; Gastli, A.; Brahim, L.B.; Touati, F.; Benammar, M. A simple method for extracting the parameters of the PV cell single-diode model. *Renew. Energy* 2017, 113, 885–894. [CrossRef]
8. Batzelis, E.I.; Papathanassiou, S.A. A Method for the Analytical Extraction of the Single-Diode PV Model Parameters. *IEEE Trans. Sustain. Energy* 2016, 7, 504–512. [CrossRef]
9. Brano, V.L.; Ciulla, G. An efficient analytical approach for obtaining a five parameters model of photovoltaic modules using only reference data. *Appl. Energy* 2013, 111, 894–903. [CrossRef]
10. Louzazni, M.; Aroudam, E.H. An analytical mathematical modeling to extract the parameters of solar cell from implicit equation to explicit form. *Appl. Sol. Energy* 2015, 51, 165–171. [CrossRef]
11. Kumar, G.; Panchal, A.K. A non-iterative technique for determination of solar cell parameters from the light generated I-V characteristic. *J. Appl. Phys.* 2013, 114, 84903. [CrossRef]
12. Saleem, H.; Karmalkar, S. An Analytical Method to Extract the Physical Parameters of a Solar Cell from Four Points on the Illuminated J-V Curve. *IEEE Electr. Device Lett.* 2009, 30, 349–352. [CrossRef]
13. Wang, G.; Zhao, K.; Shi, J.; Chen, W.; Zhang, H.; Yang, X.; Zhao, Y. An iterative approach for modeling photovoltaic modules without implicit equations. *Appl. Energy* 2017, 202, 189–198. [CrossRef]
14. Wolf, P.; Benda, V. Identification of PV solar cells and modules parameters by combining statistical and analytical methods. *Sol. Energy* 2013, 93, 151–157. [CrossRef]
15. Toledo, F.J.; Blanes, J.M. Analytical and quasi-explicit four arbitrary point method for extraction of solar cell single-diode model parameters. *Renew. Energy* 2016, 92, 346–356. [CrossRef]
16. Yeh, W.C.; Huang, C.L.; Lin, P.; Chen, Z.; Jiang, Y.; Sun, B. Simplex Simplified Swarm Optimization for the Efficient Optimization of Parameter Identification for Solar Cell Models. *IET Renew. Power Gener.* 2018, 12, 45–51. [CrossRef]
17. Bastidasrodriguez, J.D.; Petrone, G.; Ramospaja, C.A.; Spagnuolo, G. A genetic algorithm for identifying the single diode model parameters of a photovoltaic panel. *Math. Comput. Simul.* 2017, 131, 38–54. [CrossRef]
18. El-Naggar, K.M.; Alrashidi, M.R.; Alhajri, M.F.; Al-Othman, A.K. Simulated Annealing algorithm for photovoltaic parameters identification. *Sol. Energy* 2012, 86, 266–274. [CrossRef]
19. Bana, S.; Saini, R.P. Identification of unknown parameters of a single diode photovoltaic model using particle swarm optimization with binary constraints. *Renew. Energy* 2017, 101, 1299–1310. [CrossRef]

20. Nunes, H.G.G.; Pombo, J.A.N.; Mariano, S.J.P.S.; Calado, M.R.A.; Souza, J.A.M.F. A new high performance method for determining the parameters of PV cells and modules based on guaranteed convergence particle swarm optimization. *Appl. Energy* **2018**, *211*, 774–791. [CrossRef]
21. Ishaque, K.; Salam, Z.; Mekhilef, S.; Shamsudin, A. Parameter extraction of solar photovoltaic modules using penalty-based differential evolution. *Appl. Energy* **2012**, *99*, 297–308. [CrossRef]
22. Chellaswamy, C.; Ramesh, R. Parameter extraction of solar cell models based on adaptive differential evolution algorithm. *Renew. Energy* **2016**, *97*, 823–837. [CrossRef]
23. Jiang, L.L.; Maskell, D.L.; Patra, J.C. Parameter estimation of solar cells and modules using an improved adaptive differential evolution algorithm. *Appl. Energy* **2013**, *112*, 185–193. [CrossRef]
24. Askarzadeh, A.; Rezazadeh, A. Artificial bee swarm optimization algorithm for parameters identification of solar cell models. *Appl. Energy* **2013**, *102*, 943–949. [CrossRef]
25. Chen, X.; Yu, K.; Du, W.; Zhao, W.; Liu, G. Parameters identification of solar cell models using generalized oppositional teaching learning based optimization. *Energy* **2016**, *99*, 170–180. [CrossRef]
26. Chen, X.; Xu, B.; Mei, C.; Ding, Y.; Li, K. Teaching–learning–based artificial bee colony for solar photovoltaic parameter estimation. *Appl. Energy* **2018**, *212*, 1578–1588. [CrossRef]
27. Yu, K.; Chen, X.; Wang, X.; Wang, Z. Parameters identification of photovoltaic models using self-adaptive teaching-learning-based optimization. *Energy Convers. Manag.* **2017**, *145*, 233–246. [CrossRef]
28. Yu, K.; Liang, J.J.; Qu, B.Y.; Chen, X.; Wang, H.; Yu, K. Parameters identification of photovoltaic models using an improved JAYA optimization algorithm. *Energy Convers. Manag.* **2017**, *150*, 742–753. [CrossRef]
29. Yu, K.; Liang, J.J.; Qu, B.Y.; Cheng, Z.; Wang, H. Multiple learning backtracking search algorithm for estimating parameters of photovoltaic models. *Appl. Energy* **2018**, *226*, 408–422. [CrossRef]
30. Oliva, D.; Aziz, M.A.E.; Hassanien, A.E. Parameter estimation of photovoltaic cells using an improved chaotic whale optimization algorithm. *Appl. Energy* **2017**, *200*, 141–154. [CrossRef]
31. Oliva, D.; Ewees, A.A.; Aziz, M.A.E.; Hassanien, A.E.; Cisneros, M.P. A Chaotic Improved Artificial Bee Colony for Parameter Estimation of Photovoltaic Cells. *Energies* **2017**, *10*, 865. [CrossRef]
32. Kichou, S.; Silvestre, S.; Guglielminotti, L.; Mora-López, L.; Muñoz-Cerón, E. Comparison of two PV array models for the simulation of PV systems using five different algorithms for the parameters identification. *Renew. Energy* **2016**, *99*, 270–279. [CrossRef]
33. Ma, J.; Bi, Z.; Ting, T.O.; Hao, S.; Hao, W. Comparative performance on photovoltaic model parameter identification via bio-inspired algorithms. *Sol. Energy* **2016**, *132*, 606–616. [CrossRef]
34. Niu, Q.; Zhang, L.; Li, K. A biogeography-based optimization algorithm with mutation strategies for model parameter estimation of solar and fuel cells. *Energy Convers. Manag.* **2014**, *86*, 1173–1185. [CrossRef]
35. Askarzadeh, A.; Rezazadeh, A. Parameter identification for solar cell models using harmony search-based algorithms. *Sol. Energy* **2012**, *86*, 3241–3249. [CrossRef]
36. Valdivia-González, A.; Zaldívar, D.; Cuevas, E.; Pérez-Cisneros, M.; Fausto, F.; González, A. A Chaos-Embedded Gravitational Search Algorithm for the Identification of Electrical Parameters of Photovoltaic Cells. *Energies* **2017**, *7*, 1052. [CrossRef]
37. Rezk, H.; Fathy, A. A novel optimal parameters identification of triple-junction solar cell based on a recently meta-heuristic water cycle algorithm. *Sol. Energy* **2017**, *157*, 778–791. [CrossRef]
38. Alam, D.F.; Yousri, D.A.; Eteiba, M.B. Flower Pollination Algorithm based solar PV parameter estimation. *Energy Convers. Manag.* **2015**, *101*, 410–422. [CrossRef]
39. Ali, E.E.; El-Hameed, M.A.; El-Fergany, A.A.; El-Arini, M.M. Parameter extraction of photovoltaic generating units using multi-verse optimizer. *Sustain. Energy Technol. Assess.* **2016**, *17*, 68–76. [CrossRef]
40. Yuan, X.; He, Y.; Liu, L. Parameter extraction of solar cell models using chaotic asexual reproduction optimization. *Neural Comput. Appl.* **2015**, *26*, 1227–1239. [CrossRef]
41. Babu, T.S.; Ram, J.P.; Sangeetha, K.; Laudani, A.; Rajasekar, N. Parameter extraction of two diode solar PV model using Fireworks algorithm. *Sol. Energy* **2016**, *140*, 265–276. [CrossRef]
42. Allam, D.; Yousri, D.A.; Eteiba, M.B. Parameters extraction of the three diode model for the multi-crystalline solar cell/module using Moth-Flame Optimization Algorithm. *Energy Convers. Manag.* **2016**, *123*, 535–548. [CrossRef]
43. Fathy, A.; Rezk, H. Parameter estimation of photovoltaic system using imperialist competitive algorithm. *Renew. Energy* **2017**, *111*, 307–320. [CrossRef]

44. Askarzadeh, A.; Rezazadeh, A. Extraction of maximum power point in solar cells using bird mating optimizer-based parameters identification approach. *Sol. Energy* **2013**, *90*, 123–133. [CrossRef]
45. Alhajri, M.F.; El-Naggar, K.M.; Alrashidi, M.R.; Al-Othman, A.K. Optimal extraction of solar cell parameters using pattern search. *Renew. Energy* **2012**, *44*, 238–245. [CrossRef]
46. Xiong, G.; Zhang, J.; Shi, D.; He, Y. Parameter extraction of solar photovoltaic models using an improved whale optimization algorithm. *Energy Convers. Manag.* **2018**, *174*, 388–405. [CrossRef]
47. Wolpert, D.H.; Macready, W.G. No free lunch theorems for optimization. *IEEE Trans. Evol. Comput.* **1997**, *1*, 67–82. [CrossRef]
48. Cheng, M.Y.; Prayogo, D. Symbiotic Organisms Search: A new metaheuristic optimization algorithm. *Comput. Struct.* **2014**, *139*, 98–112. [CrossRef]
49. Saha, S.; Mukherjee, V. A novel chaos-integrated symbiotic organisms search algorithm for global optimization. *Soft Comput.* **2018**, *22*, 3797–3816. [CrossRef]
50. Panda, A.; Pani, S. A Symbiotic Organisms Search algorithm with adaptive penalty function to solve multi-objective constrained optimization problems. *Appl. Soft Comput.* **2016**, *46*, 344–360. [CrossRef]
51. Shongwe, S.; Hanif, M. Comparative Analysis of Different Single-Diode PV Modeling Methods. *IEEE J. Photovolt.* **2015**, *5*, 938–946. [CrossRef]
52. Humada, A.M.; Hojabri, M.; Mekhilef, S.; Hamada, H.M. Solar cell parameters extraction based on single and double-diode models: A review. *Renew. Sustain. Energy Rev.* **2016**, *56*, 494–509. [CrossRef]
53. Deihimi, M.H.; Naghizadeh, R.A.; Meyabadi, A.F. Systematic derivation of parameters of one exponential model for photovoltaic modules using numerical information of data sheet. *Renew. Energy* **2016**, *87*, 676–685. [CrossRef]
54. Jordehi, A.R. Maximum power point tracking in photovoltaic (PV) systems: A review of different approaches. *Renew. Sustain. Energy Rev.* **2016**, *65*, 1127–1138. [CrossRef]
55. Ishaque, K.; Salam, Z.; Taheri, H. Simple, fast and accurate two-diode model for photovoltaic modules. *Sol. Energy Mater. Sol. Cells* **2011**, *95*, 586–594. [CrossRef]
56. Awadallah, M.A. Variations of the bacterial foraging algorithm for the extraction of PV module parameters from nameplate data. *Energy Convers. Manag.* **2016**, *113*, 312–320. [CrossRef]
57. Chen, Z.; Wu, L.; Cheng, S.; Lin, P.; Wu, Y.; Lin, W. Intelligent fault diagnosis of photovoltaic arrays based on optimized kernel extreme learning machine and I-V characteristics. *Appl. Energy* **2017**, *204*, 912–931. [CrossRef]
58. Easwarakhanthan, T.; Bottin, J.; Bouhouch, I.; Boutrit, C. Nonlinear minimization algorithm for determining the solar cell parameters with microcomputers. *Int. J. Sol. Energy* **1986**, *4*, 1–12. [CrossRef]
59. Wu, G. Across neighborhood search for numerical optimization. *Inform. Sci.* **2016**, *329*, 597–618. [CrossRef]
60. Chen, X.; Tianfield, H.; Mei, C.; Du, W.; Liu, G. Biogeography-based learning particle swarm optimization. *Soft Comput.* **2017**, *21*, 7519–7541. [CrossRef]
61. Cheng, R.; Jin, Y. A competitive swarm optimizer for large scale optimization. *IEEE Trans. Cybern.* **2015**, *42*, 191–204. [CrossRef] [PubMed]
62. Farah, A.; Guesmi, T.; Abdallah, H.H.; Ouali, A. A novel chaotic teaching-learning-based optimization algorithm for multi-machine power system stabilizers design problem. *Int. J. Electr. Power* **2016**, *77*, 197–209. [CrossRef]
63. Ling, Y.; Zhou, Y.; Luo, Q. Lévy flight trajectory-based whale optimization algorithm for global optimization. *IEEE Access* **2017**, *5*, 6168–6186. [CrossRef]

© 2018 by the authors. Licensee MDPI, Basel, Switzerland. This article is an open access article distributed under the terms and conditions of the Creative Commons Attribution (CC BY) license (http://creativecommons.org/licenses/by/4.0/).

Article

An Evolutionary-Based MPPT Algorithm for Photovoltaic Systems under Dynamic Partial Shading

Alberto Dolara *, Francesco Grimaccia, Marco Mussetta, Emanuele Ogliari and Sonia Leva

Department of Energy, Politecnico di Milano, via La Masa 34, 20156 Milano, Italy; francesco.grimaccia@polimi.it (F.G.); marco.mussetta@polimi.it (M.M.); emanuelegiovanni.ogliari@polimi.it (E.O.); sonia.leva@polimi.it (S.L.)
* Correspondence: alberto.dolara@polimi.it; Tel.: +39-02-2399-3829

Received: 8 March 2018; Accepted: 29 March 2018; Published: 4 April 2018

Abstract: The increase of renewable energy usage in the last two decades, in particular photovoltaic (PV) systems, has opened up different solar plant configurations that need to operate and properly perform in terms of efficient power transfer with respect to all of the involved components, such as inverters, grid interface, storage, and other electrical loads. In such applications, the power characteristics of the plant modules all together represent the main components that are responsible for power extraction, depending on both external and internal factors. Conventional maximum power point tracking techniques may not have a proper conversion efficiency under particular external dynamic conditions. This paper proposes an evolutionary-based maximum power point tracking algorithm suitable to operate under dynamic partial shading conditions and compares its performance with classical maximum power point tracking methods in order to evaluate their conversion efficiency in partial shading scenarios with relevant and dynamic changes in the environmental conditions. Simulations taking into account the different dynamic shading conditions were carried out to prove the effectiveness and limitations of the proposed approach with respect to classical algorithms.

Keywords: photovoltaics; MPPT algorithm; evolutionary algorithms; particle swarm optimization

1. Introduction

Solar energy is one of most reliable, cleanest, and easy to harvest energy sources among all renewables. Solar energy can be converted into other forms of energy, mostly electricity and heat. Solar cells convert sunlight directly into electricity, while concentrating solar power systems use mirrors to concentrate the energy from the sun to heat a working fluid that drives traditional steam turbines connected to an electrical power generator. Over the last two decades, worldwide solar photovoltaic (PV) installations have exponentially grown and their cost has significantly reduced due to the improvements in technology and the economies of scale. Besides direct economic and environmental advantages, the use of solar energy produces other indirect environmental and social benefits, such as the stabilization of degraded land, the increasing of energy independence, new job opportunities, acceleration of remote rural areas electrification, and improved quality of life in developing countries [1]. The environmental issues related to the installation of solar power facilities have been comprehensively addressed in [2]: more than thirty environmental impacts falling in the fields of land use intensity, human health and wellbeing, plant and animal life, geo-hydrological resources, and climate change. In this context, the integration of solar energy into islanded micro-grids can play a key role both in decreasing the dependence on fossil fuels in energy intensive applications, for example in the desalination of plants [3], and to speed up access to electricity in developing countries [4].

The high penetration of renewables in the power grid presents several challenges, mainly related to the uncertainty in generation output due to weather fluctuations. In recent years, several Computational

Intelligence (CI) techniques have been developed for renewable energies [5], with the aim of helping grid operators to better manage the electric balance between power demand and supply. In particular, the efficiency of a whole PV system depends on many factors at the same time: the efficiency of each single PV module [6,7], mismatch losses in the PV generator [8,9], losses in the wiring, switches, transformers and power converters, and the extraction efficiency of the Maximum Power Point Tracking (MPPT) algorithm. Moreover, the efficiency of the hardware components is neither easily or cheaply improved, but the improvement of the MPPT with new control algorithms allows for increasing power generation performance. In this context, CI techniques can be successfully employed to tackle various problems in system optimization and power extraction strategy [10]. In this light, the Particle Swarm Optimization (PSO) is a well-known evolutionary algorithm based on a model of group interaction between independent agents (particles) that uses social behavior (or *swarm knowledge*) in order to find a global maximum or minimum of a specific fitness function [11]. This computational technique adopts a biological-like approach, and takes its rationale from the emulation of social behaviors such as those related to bird flocking and fish schooling. The PSO algorithm explores the problem domain with moving particles represented by a set of coordinates in the N-dimensional solution space, which represents a specific problem solution corresponding to a particular value of the objective function.

This work proposes a novel MPPT algorithm based on the PSO evolutionary approach and compares the results with traditional MPPT methods, previously tested by the authors [12], with the aim of increasing the overall conversion efficiency in particular conditions characterized by relevant changes and high dynamism with an affordable computational burden. This optimization of solar power extraction requires operating the photovoltaic generator at its maximum efficiency and can naturally benefit evolutionary computation. Numerous research studies on MPPT for solar PV systems have been proposed in last few decades and the methods developed so far can be broadly classified into conventional versus soft computing techniques [13].

Conventional MPPT algorithms are based on search algorithms designed to find the maximum output in different external conditions [14], but only under uniform irradiation. These methods generally fail when partial shading occurs on the PV generator, showing poor convergence, slow tracking speed, and often high steady state oscillations [15,16]. Among the conventional MPPT methods, the most popular are Perturb and Observe (P&O) [17] and Incremental Conductance (IC) [18].

In order to overcome the drawbacks of a specific conventional MPPT technique and to gain better performance, various methods are combined to set up MPPT hybrid algorithms [19,20]. Soft computing and evolutionary algorithms can offer some advantages, such as their ability to handle non linearity, a wide exploration in search space, and coherent skill to reach global optimal regions [15]. Moreover, they can be combined with conventional MPPT techniques [21,22] in order to further improve their efficiency.

In literature, other comparative studies can be found among different MPPT algorithms on PV modules under dynamic shading conditions [23,24]. Since partial shading on PV modules may result in power curves with multiple peaks and multiple local minima and maxima [25], the main advantage of using evolutionary computation techniques results in their acting as an effective global optimizer. In this context, the PSO-based MPPT algorithm here proposed appears to be a good tradeoff between simplicity in implementation and accuracy in the tracking of the Global Maximum Power Point (GMPP). This work aims also to compare this novel algorithm based on a bio-inspired approach with traditional MPPT methods in order to evaluate efficiency under partial and dynamic shading conditions.

The paper is structured as follows: Section 2 describes critical issues of MPPT algorithms under partial shading conditions and describes, step by step, the proposed MPPT algorithm based on the PSO evolutionary approach. Section 3 reports the simulated test case scenario with related applied shading conditions. Section 4 shows the numerical results and the preliminary experimental tests of the proposed algorithm in comparison with other classical MPPT algorithms with respect to different shading objects and various dynamics. Section 5 draws conclusions based on the results.

2. Evolutionary-Based MPPT Algorithm for PV System Power Extraction

2.1. MPPTs and PV System Issues Under Dynamic Partial Shading

PV generators are highly dependent on environmental conditions such as solar irradiance, temperature, and particular shading conditions. The natural variation of such environmental parameters combined with the nonlinear characteristics of the solar cells make power extraction optimization a critical task. To obtain the maximum conversion efficiency, the PV generator must operate at the higher point of its power-voltage curve, also called the Maximum Power Point (MPP). This control is known as the Maximum Power Point Tracking, and it is the typical operating condition of PV generators both in grid-connected and off-grid PV systems. Only in very special cases, the PV generator is regulated at a lower power than the MPP with the algorithm of Regulated Power Point Tracking (RPPT), as in the case of islanded hybrid micro-grids where the power production exceeds the load demand and storage systems are not available [4]. Various MPPT techniques have been developed over the last years that follow both direct and indirect approaches.

When particular shading conditions occur on a PV generator, one or more bypass diodes can be forward biased to prevent localized power dissipation at the shaded cells. The bypass diodes forward or reverse bias state under partial shading depends on the PV generator output voltage, resulting in modified P-V curves with multiple local maxima. Thus, the MPPT algorithm running in the control unit must be adapted or modified when designing an algorithm suitable to operate in the case of partial shading and rapidly changing environmental conditions.

Traditional Perturb and Observe methods (P&Os) simply compare the actual PV generator voltage and output power with respect to the previous ones, adapting the operating points accordingly to reach the MPP condition. However, when rapid environmental changes occur, even the most sophisticated variant of P&Os may lose efficiency, since these hill climbing techniques may not always reach the GMPP and the operating point may get stuck in a local maxima. Thus, it is possible to conceive a family of PSO-based MPPT algorithms, with a variable population size, in order to better find the proper MPP. On the other hand, by increasing the number of particles, the computational time increases and the conversion efficiency decreases. Thus, in the context of PSO-based MPPT algorithms it is better to keep the population size as small as possible. Moreover, in this particular application characterized by dynamic partial shading, a time-varying fitness function has to be considered. Among all other evolutionary techniques, Particle Swarm Optimization is characterized by easy implementation. In this particular context, the exploitation feature of this algorithm can guarantee to reach the MPP without oscillations. Knowing a priori the number of potential local maxima related to the specific system and number of diodes, it is possible to implement a simple PSO-based algorithm with a relatively small number of agents, from three to five in this case, in order to explore the one-dimensional solution space. The next subsection describes in detail the implementation of the proposed PSO-based MPPT algorithm.

2.2. PSO-Based Algorithm Implementation

PSO is an evolutionary algorithm based on the modelling of group interaction between independent *agents*, also called *particles*, that share information about their respective search process in order to find a global maximum or minimum of a specific fitness function. Each particle represents a candidate solution and their movement in the search domain depends both on its own previous best position and the previous best position attained among all the particles. This behavior is mathematically expressed by two equations that define the velocity, v_i, and the position, x_i, of the i-th agent at the k-th step of the searching process:

$$v_i^{k+1} = w \cdot v_i^k + c_1 \cdot r_1 \cdot \left(p_{best,i} - x_i^k\right) + c_2 \cdot r_2 \cdot \left(g_{best} - x_i^k\right) \tag{1}$$

$$x_i^{k+1} = x_i^k + v_i^{k+1} \tag{2}$$

where w is the inertia weight, c_1 and c_2 are the acceleration coefficients, r_1 and r_2 are random numbers between 0 and 1, $p_{best,i}$ is the best position attained by the i-th particle and g_{best} is the best position attained among all the particles.

It was found that a basic problem in the application of the PSO for the MPPT system is in its random nature [16]. Very low values of r_1 and r_2 result in a very low velocity, thus a large number of iteration is required to reach the solution. On the other hand, too large changes in the velocity might cause the particles to move far from the neighborhood of the global maximum, opening up the possibility of converging to a local maximum instead of the global one. Moreover, in a partially shaded PV generator, the distance between the two consecutive peaks in the P-V curve is quite constant, about 80% of the open voltage of the string of PV cells connected in parallel with a bypass diode.

The goal of a MPPT algorithm is to maximize the PV output power by adjusting the duty cycle of the power converter dedicated to this function. The search of the global maximum using PSO is simple due to the fact that for this problem only a single dimensional search space is needed: the particles represent the duty cycle values and the PV generated power is the fitness function. The PSO-based MPPT algorithm presented in this work is based on a more deterministic structure, removing the random factors in (1). The velocity is limited to comply with the distance between the two peaks by tuning the coefficients c_1 and c_2. The search of the global maximum in the PSO-based MPPT algorithm is mathematically expressed by two equations that define the variation of the duty cycle, Δd_i, and the duty cycle, d_i, corresponding to the i-th agent at the k-th step of the searching process:

$$\Delta d_i^{k+1} = w \cdot d_i^k + c_1 \cdot \left(d_{best,i} - d_i^k \right) + c_2 \cdot \left(D_{best} - d_i^k \right) \tag{3}$$

$$d_i^{k+1} = d_i^k + \Delta d_i^{k+1} \tag{4}$$

where w is the inertia weight, c_1 and c_2 are the acceleration coefficients, $d_{best,i}$ is the duty cycle corresponding to the maximum PV output power detected by the i-th particle and D_{best} is the duty cycle corresponding to the maximum PV output power detected among all the particles.

Figures 1 and 2 show the flowcharts of the proposed PSO-based MPPT algorithm; they represent the program of the MPPT controller that runs at every n-th time interval.

The proposed PSO-based MPPT algorithm is organized in the following seven steps.

Step 1: *Activation of the MPPT algorithm.* First, the MPPT controller checks whether it is necessary to activate the search for a new GMPP by comparing the actual PV power, $P_{PV}(n)$, with the previously recorded maximum power, P_{MPP}. The absolute value of the difference between these powers over a threshold triggers the activation of the search for a new GMPP.

Step 2: *PSO Initialization.* This step is the beginning of each GMPP searching process. The initial positions of the particles, namely a first solution vector of duty cycles with N_p elements, are calculated: they are linearly spaced between the minimum and maximum duty cycle to cover the whole search space. All of the variables that contain information related to a previous GMPP search are reset. Then, the algorithm transmits the first duty cycle to the power converter. The change of duty cycle causes a transient of the electrical system; the new steady state condition will be evaluated at the next time interval.

Step 3: *Fitness Evaluation and Update Individual Best Data.* The fitness value $P_{PV}(n)$ (actual PV output power) of the i-th particle, is calculated. Both the best individual position, $d_{best,i}$, and the best fitness, $P_{PVbest,i}$, are updated in case the fitness value is greater than the best fitness. If the particle that has just been evaluated is not the last one, the algorithm transmits the next duty cycle to the power converter, whose fitness function will be evaluated at the next time interval. Otherwise, global best data will be updated and operations after each k-th iteration will be performed.

Step 4: *Update Global Best Data and End-of-Iteration checks.* This step is the end of each k-th iteration. Both the global best position, D_{best}, and the global best fitness, $P_{PV,Gbest}$, are updated in case the maximum of the best fitness values of particles is greater than the global best fitness. A counter, $CounterG_{best}$, is set to 1 at every global best position update, otherwise it is incremented.

Step 5: *Convergence Determination and Reset Criterion.* The convergence determination proposed in this work is based on the number of iterations without the update of D_{best}: the convergence is reached in case a new D_{best} is not found in the last N_{Gbest} iterations. Moreover, a maximum number of iterations, N_{Iter}, is allowed to reach convergence. In case the convergence is not met within the number of allowed iterations, the algorithm has to update the velocity and position of each particle and has to perform another search iteration (move to Step 6). In case the convergence is met within the number of allowed iterations, the algorithm transmits D_{best} to the power converter to check the solution (move to Step 7). If the maximum number of allowed iterations is reached without convergence, the search for the GMPP has to be repeated (move to Step 2).

Step 6: *Update Velocity and Position of Each Particle.* After all the particles are evaluated and convergence is not achieved, the velocity and the position of each particle have to be updated by using Equations (3) and (4). Then, the algorithm transmits the first duty cycles to the power converter. At the next time interval, the algorithm will move to Step 3.

Step 7: *Check the GMPP.* The duty cycle of the power converter is D_{best}; the actual PV output power, $P_{PV}(n)$, is compared with the global best fitness, $P_{PV,Gbest}$. The GMPP is reached if the absolute value of the difference between these powers is below a threshold. This check is necessary in case of dynamic partial shading, when the P-V curve changes significantly during the GMPP search process. Under these conditions, several fitness functions are sampled during the GMPP search process, making the information of $d_{best,i}$ and D_{best} totally useless for tracking the actual GMPP. In case the result of the GMPP check is not successful, a new full scan is necessary and the algorithm immediately moves to Step 2. Otherwise, it is assumed that the GMPP has been reached and the power converter will be operated with the duty cycle D_{best} until a change in the environmental conditions, namely a change in the PV output power, triggers a new scan.

Figure 3 shows the timing of the proposed PSO-based MPPT algorithm. In the example of Figure 3, a decrease of PV output power triggers a new search of the GMPP and at the end of the third iteration an update of D_{best}, $P_{PV,Gbest}$ occurs; N_{Gbest} is 4 and N_{Iter} is 8. Iterations from 3 to 6 are characterized by the same D_{best}, thus the end of the sixth iteration triggers the check of the solution.

The time required for the search of the GMPP, T, is reported in Equation (5), and it depends on the number of particles to be evaluated in each iteration, N_p, the number of iterations required to meet the convergence criterion, and the sampling time T_s. In order to avoid the measurement of the PV output power during transients, T_s has to be larger than the power converter settling time. It is easy to prove that the search process requires a quite long time interval in which the PV generator works at different power values, even far from the GMPP.

$$T = (N_p \cdot CounterITER + 1) \cdot T_s \qquad (5)$$

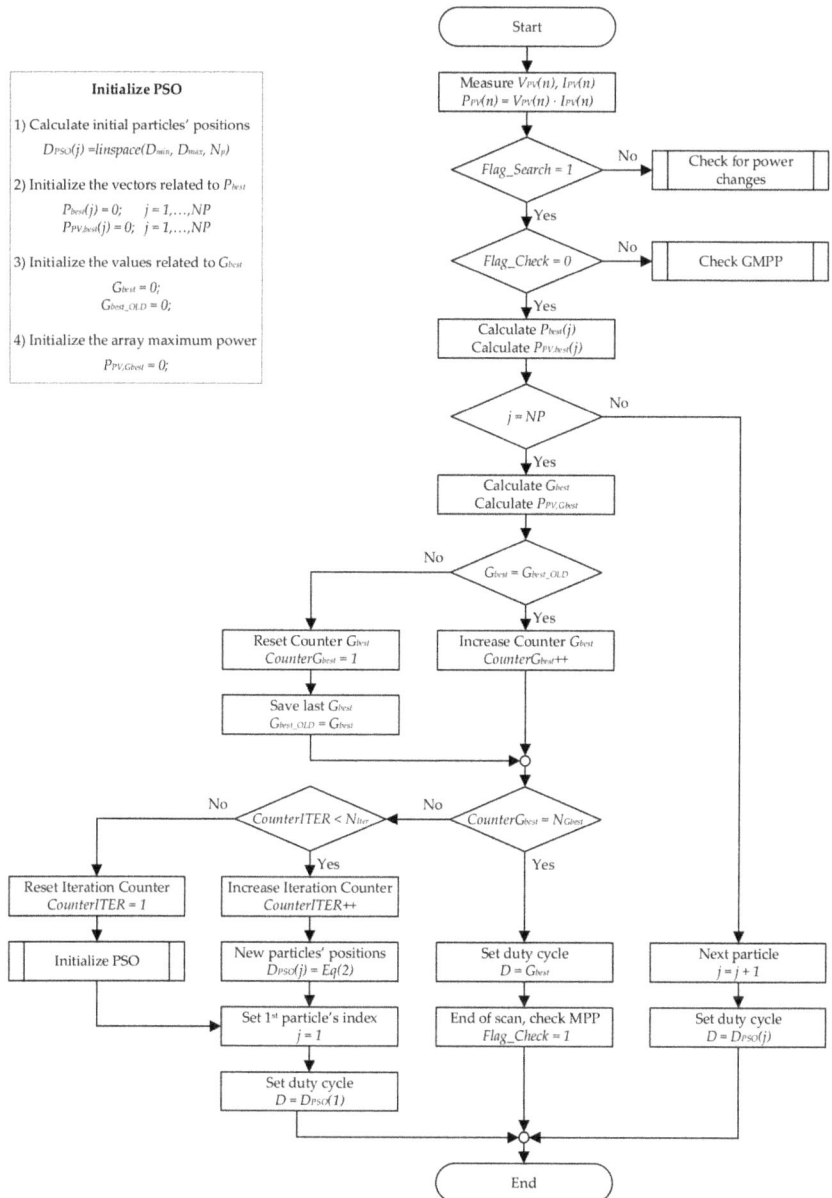

Figure 1. Flowchart of Particle Swarm Optimization (PSO)-based Maximum Power Point (MPP) tracking method.

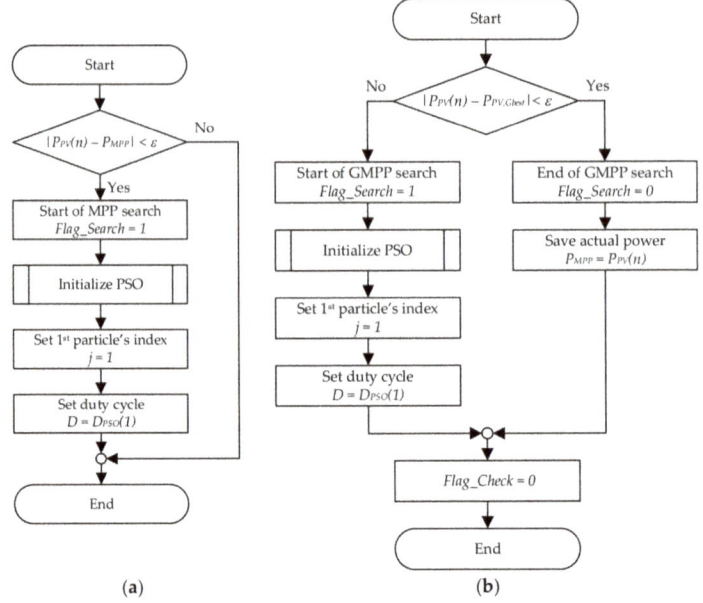

Figure 2. Flowchart of the functions (**a**) "Check for power changes" and (**b**) "Check GMPP".

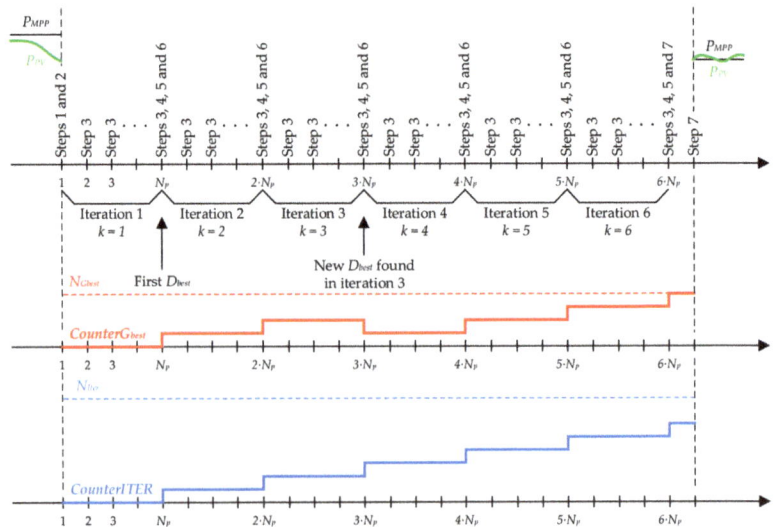

Figure 3. Timing of the proposed PSO-based Maximum Power Point Tracking (MPPT) algorithm.

3. Test Case Modeling

3.1. Test Case PV System Model

The test case PV system taken into account for simulating the MPPT algorithms is a simple off grid PV system that consists of a PV module, a boost DC/DC converter, a load resistor, and a MPPT controller. Figure 4 shows the diagram of the simulated off grid PV system. This model represents the MPPT test bench available at the SolarTechLab of the Department of Energy at Politecnico di Milano [24], which was used for a preliminary experimental test campaign carried out on 27 February and 3 March 2018. The MPPT algorithm controls the PV generator, regardless of the kind and the complexity of the PV system. Since the goal of this work is to investigate, by means of simulations, the capability of the MPPT controller to find the GMPP under partial and dynamic shading conditions, this simple off-grid PV system represents a general case suitable for this kind of research.

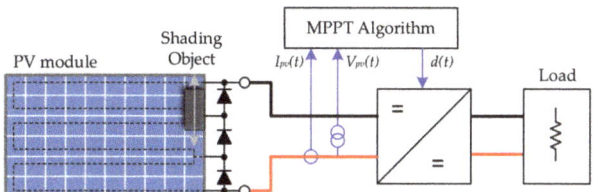

Figure 4. Diagram of the simulated off grid photovoltaic (PV) system.

The PV module taken into account in this work is a monocrystalline S19.285 produced by Aleo Solar, whose ratings are reported in Table 1. It is made of 60 cells connected in series and three bypass diodes parallel connected to groups of 20 cells, which divide the PV module in three equal electrical sections. An active non-linear and time-varying element models the PV module. The PV module I-V curve is calculated combining the I-V curves of each cell and the I-V curves of each bypass diode in order to comply with the series and parallel electrical connection constraints. The I-V curve of each cell was calculated by using the five parameter mathematical model [26], whose equivalent circuit is reported in Figure 5. The five parameters that characterize this model are the light generated current (I_{PV}), the leakage or reverse saturation current (I_0), the diode quality factor (n), the series resistance (R_s) and the shunt resistance (R_{sh}). Referring to the equivalent electric circuit in Figure 5, the I-V curve of a PV cell can be expressed based on Kirchhoff's current law, Ohm's law, and the Shockley diode equation:

$$I(G, T_C) = I_{PV}(G, T_C) - I_0(T_C) \cdot \left(e^{\frac{V + R_s \cdot I(G,T_C)}{n \cdot V_t}} - 1 \right) - \frac{V + R_s \cdot I(G, T_C)}{R_{SH}(G)} \quad (6)$$

The light generated current as a function of irradiance, G, and cell temperature, T_C, can be expressed as:

$$I_{PV}(G, T_C) = I_{PV,ref} \cdot \frac{G}{G_{ref}} \cdot \frac{S_{unshaded}}{S_{cell}} \cdot \left(1 + \alpha_{ISC} \cdot (T_C - T_{ref}) \right) \quad (7)$$

where the subscript *ref* stands for reference conditions, G is the irradiance, T_C is the cell temperature and α_{ISC} is the temperature coefficient for short-circuit current. In most cases, reference values are measured at standard test conditions (STC), that is with G_{ref} equal to 1000 W/m², cell temperature equal to 25 °C and Air Mass equal to 1.5. In case of partial shading on a cell, its light generated current is directly proportional to the ratio of the unshaded area of the cell, $S_{unshaded}$, and its total surface, S_{cell}.

The unshaded area of each cell is calculated according to the shape of the shading object and its speed and position. The diode reverse saturation current can be expressed as:

$$I_0(T_C) = I_{0,ref} \cdot \left(\frac{T_C}{T_{ref}}\right)^3 \cdot e^{\frac{E_g(T_{REF})}{n \cdot k \cdot T_{ref}} - \frac{E_g(T_C)}{n \cdot k \cdot T_C}} \tag{8}$$

where k is the Boltzmann constant ($8.6173324 \cdot 10^{-5}$ eV·K^{-1}) and E_g is the bandgap energy of the silicon, that is temperature dependent and it is given, in eV, as:

$$E_g(T) = 1.17 - 4.73 \cdot 10^{-4} \frac{T^2}{T + 636} \tag{9}$$

The shunt resistance changes with absorbed solar radiation, and different equations can be found in literature. In this work, inversely proportional dependence of shunt resistance with irradiance is taken into account.

$$R_{SH}(G) = R_{SH,ref} \cdot \frac{G_{ref}}{G} \tag{10}$$

Variation of the ideality factor of the cell and the series resistance with irradiance and cell temperature are neglected.

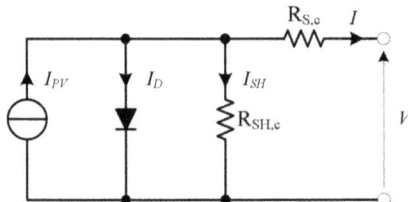

Figure 5. Diagram of the simulated off grid PV system.

Table 1. PV module electrical data.

Electrical Data		STC [1]	NOCT [2]
Rated Power	P_{MPP} (W)	285	208
Rated Voltage	V_{MPP} (V)	31.3	28.4
Rated Current	I_{MPP} (A)	9.10	7.33
Open-Circuit Voltage	V_{OC} (V)	39.2	36.1
Short-Circuit Current	I_{SC} (A)	9.73	7.87

[1] Electrical values measured under Standard Test Conditions: 1000 W/m^2, cell temperature 25 °C, AM 1.5;
[2] Electrical values measured under Nominal Operating Conditions of cells: 800 W/m^2, ambient temperature 20 °C, AM 1.5, wind speed 1 m/s, NOCT: 48 °C (nominal operating cell temperature).

The MPPT controller regulates the duty cycle according to the MPPT algorithm. As the MPPT controller always operates with the average PV voltage and current measured in the steady-state, the dynamics of the DC/DC converter, namely the voltage and current high frequency ripples, can be neglected. Moreover, MPPT controller regulates DC/DC converter input voltage and current regardless of its internal losses. In this context, the boost DC-DC converter is represented by its low-frequency ideal model, consisting of an ideal transformer, whose transformation ratio is:

$$\frac{V_{out}(t)}{V_{PV}(t)} = \frac{1}{1 - d(t)} \tag{11}$$

where $d(t)$ is the duty cycle, $V_{out}(t)$ is the output voltage applied to the load resistor, and $V_{pv}(t)$ is the input voltage applied to the PV module. According to the MPPT test facility available in the SolarTechLab, the resistance of the load resistor, R_{load}, is 37.5 Ω. At the PV module terminals, the load and the DC-DC converter can be replaced by an equivalent time-varying resistor, $R_{eq}(t)$.

$$R_{eq}(t) = (1 - d(t))^2 \cdot R_{load} \qquad (12)$$

The solution of the electrical model is the intersection between the I-V curve representing the actual electrical behavior of the PV module and the load line representing the resistor.

3.2. Test Case Shading Scenarios Model

The simulated dynamic partial shading scenarios take into account the passage of a moving object close to the PV module, obstructing direct radiation and most of the diffuse radiation. In the model, it has been assumed that the irradiance is nil on the shaded area of the PV module. Four shading shapes are taken into account: a rectangular shape and three trapezoidal shapes. For each shape, three shading objects that are equal in height and differ only in length are considered, as summarized in Table 2. Taking the length of the solar cell edge as the unit of measurement, the dimensions of each shading object are 2 × 1, 3 × 1, and 4 × 1. The shading scenarios with rectangular shapes were also tested during the preliminary test campaign.

The shading object moves at a constant speed (4 cm/s) along the short edge of the PV module, as shown in Figure 6. The motion path is set between two points, A and B, respectively; the PV module is placed in the middle of the path. A motion cycle consists of a forward motion from A to B, a waiting time, and a backward motion from B to A.

Table 2. Shading objects.

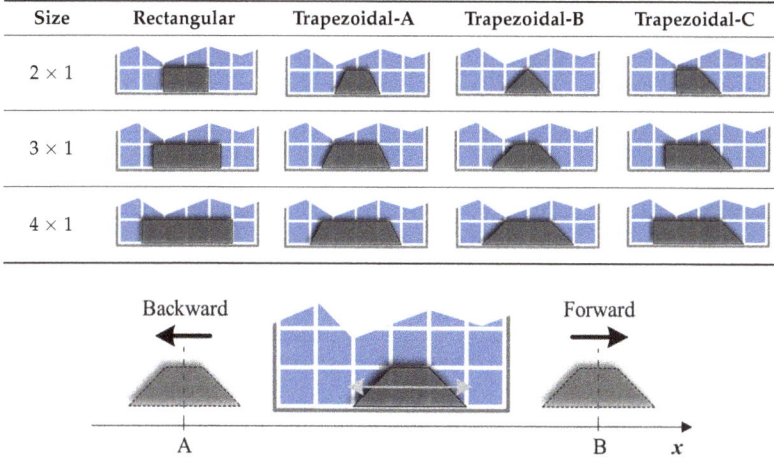

Figure 6. Diagram of the motion cycle: the shading object is moved forward and backward in one cycle.

The shapes and size of the shading objects have been defined to consider different dynamic partial shading cases, in terms of the number of completely and partially shaded cells at the same time, the rate of variation of the shaded area of PV cells, and the duration of the shading process. In a real world scenario, the shading process of a photovoltaic generator can be either deterministic or random. A deterministic partial shading process occurs in fixed PV installations; it is related to the motion of the sun in the sky and the fixed objects nearby to the PV generator [27], like trees, buildings, poles, etc., that can cause partial occlusion from the sun in some hours of the day. A deterministic partial shading process is usually a very slow process. A random partial shading process occurs both in fixed PV installations, due to the passage of clouds, and in stand-alone and moving applications, (e.g., cars or boats) because of motion. Fast dynamics characterize these types of shading. A random partial shading process is usually quite a fast process. The moving objects and their motion speed taken into account in this work allows us to simulate the shading processes that fall in the last case.

4. Test Case Simulations and Results

The test case simulations have been designed to verify the performance of the proposed PSO-based MPPT algorithm in different dynamic shading conditions and with different settings. The results obtained with the proposed PSO-based MPPT algorithm are compared with those generated by the use of widely used MPPT algorithms in the same conditions of dynamical shading. The performance of the proposed PSO-based MPPT algorithm as a function of the number of particles (population size) is investigated.

The following MPPT algorithms have been simulated for each shading scenario:

1. Perturb and Observe (P&O—A);
2. Variable Step Perturb and Observe (P&O—B);
3. Three Point Weight Comparison (P&O—C);
4. Incremental Conductance (IC);
5. Constant Voltage (CV);
6. Open Voltage (OV);
7. Short Current Pulse (SC);
8. PSO-based MPPT algorithm with 3 particles (PSO—3p);
9. PSO-based MPPT algorithm with 4 particles (PSO—4p);
10. PSO-based MPPT algorithm with 5 particles (PSO—5p).

For each shading object defined above and for each MPPT algorithm, two minutes of operation of the off grid PV system are simulated. The irradiance and cell temperature are constant during the whole simulation, equal to 800 W/m^2 and 48 °C, respectively. Since MPPT algorithms regulate the PV generator both for irradiance changes and in case of dynamic partial shading, therefore constant irradiance is necessary to evaluate the behavior of the controller in the event of dynamic partial shading. Moreover, constant irradiance represents the environmental condition during a short time interval of a sunny day. The commonly used thermal model for predicting the temperature of PV cells in a PV module is based on the assumption that the difference between the cell and ambient temperature is proportional with the irradiance, thus constant irradiance corresponds to constant cell temperature. Partial shading on PV cells persists only for a few seconds: it is assumed that the cell temperature variation due to the passage of the moving object can be neglected.

The schedule of each partial shading simulation is organized as follows:

- At the beginning of the simulation (conventional time stamp $t = 0$ s), the PV module is unshaded and the MPP algorithm is in steady state.
- The dynamical shading produced by the object moving in forward direction starts 10 s after the beginning of the simulation (conventional time stamp $t = 10$ s). The dynamical partial shading condition ends as soon as the moving object's rear-end leaves the PV module. Dynamic partial shading lasts for 31.7 s in case of 2 × 1 shading objects, for 35.7 s in case of 3 × 1 shading objects and for 39.7 s in case of 4 × 1 shading objects;
- The dynamical shading produced by the object moving in backward direction starts at the conventional time stamp $t = 70$ s. As in the case of forward motion, the dynamical partial shading lasts from 31.7 s to 39.7 s, depending on the length of shading objects.
- The simulation ends at the conventional time stamp $t = 120$ s.

Simulations have been performed by using a Matlab script that implements the calculation of the position of the moving object and the resulting shaded area of each cell, the electric model of the PV system, and the MPPT algorithm. The two minutes of operation of the off-grid PV system have been discretized using a sampling time of 50 ms. For each time sample:

1. the position of the shading object and the resulting unshaded area of each cell of the PV module are calculated;
2. the I–V curves of each PV cell and the I–V curve of the PV module are calculated and the electrical circuit is solved, namely $V_{PV}(n)$ and $I_{PV}(n)$ are calculated and $P_{PV}(n)$ is derived;
3. one cycle of the MPPT algorithm is performed and the duty cycle is changed accordingly.

In order to compare the results of 120 simulation runs (10 MPPT algorithms and 12 shading scenarios) an appropriate *Efficiency* parameter of the MPPT algorithm has been defined and calculated. This efficiency parameter is represented by the ratio of the harvested energy and the maximum harvestable energy, and can be mathematically expressed by the following equations:

$$Efficiency = \frac{E}{E_{id}} \tag{13}$$

$$E = \int_{t_{start}}^{t_{end}} V_{PV}(t) \cdot I_{PV}(t) \cdot dt \tag{14}$$

$$E_{id} = \int_{t_{start}}^{t_{end}} V_{MPP}(t) \cdot I_{MPP}(t) \cdot dt \tag{15}$$

where $V_{PV}(t)$ and $I_{PV}(t)$ are the actual voltage and current at the terminals of PV generator, $V_{MPP}(t)$ and $I_{MPP}(t)$ are the coordinates of the actual GMPP.

4.1. Results: Rectangular Shape

Table 3 reports the efficiency calculated for all the dynamical shading scenarios characterized by a rectangular shape's objects; the results obtained both from simulations and from preliminary experimental tests are reported. In the latter case, the plane of array global irradiance and the PV cell temperature during each test were recorded by the weather station installed in the SolarTechLab, and the maximum harvestable energy was estimated accordingly. A stepper motor moves the carriage with the shading object at constant speed along a rail and the signals from two optical sensors on the rail, synchronized with the shadow produced by the shape on the PV module, trigger the beginning and the ending of the partial shading process [24]. Figure 7 shows, as example, the diagrams of the harvested power (solid blue line) and maximum harvestable power (dotted red line): these are obtained in case of dynamical shading caused by the rectangular 2 × 1 shading object and controlling the off grid PV system, with the best performing traditional MPPT algorithm (P&O—B) and the proposed PSO-based MPPT algorithm characterized by three, four, and five agents respectively.

The comparison with these diagrams highlight the benefits and drawbacks of the proposed PSO-based MPPT algorithm under a quite general case of a dynamic partial shading scenario. The PSO-based MPPT algorithm reacts immediately to any variation in the PV output power caused by dynamic shading, while Variable Step Perturb and Observe requires quite a long time to move to another peak, and during this process could get stuck in a specific (local) peak. The latter characteristics, that are a common feature of the hill-climbing methods, can be both a benefit or a drawback, depending on how the shade moves on the surface of the PV module. On the other hand, the PSO-based MPPT algorithm always tracks the GMPP, but the search of the GMPP requires quite a long time interval in which the PV generator is operated at different power values, even far from the GMPP, thus reducing conversion efficiency. Moreover, the PSO-based MPPT algorithm is not suitable for tracking the GMPP when its power changes during scan.

The efficiency gap between the best PSO-based MPPT algorithm and the best hill climbing technique varies from 2.71% (3 × 1, P&O—A vs. PSO—3p) to 5.95% (4 × 1, P&O—A vs. PSO—5p). This behavior is strictly related to the long time required to track the GMPP, also combined with the inadequacy of the PSO-based MPPT method when the power of the GMPP changes during the scan phase.

The comparison among the results generated by the PSO-based MPPT algorithm with a variable number of agents shows that there is no specific rule to determine the optimum number of particles. As the number of particles increases, there is a tendency to reduce the number of iterations required to track the GMPP, but each iteration requires more time to evaluate the fitness function.

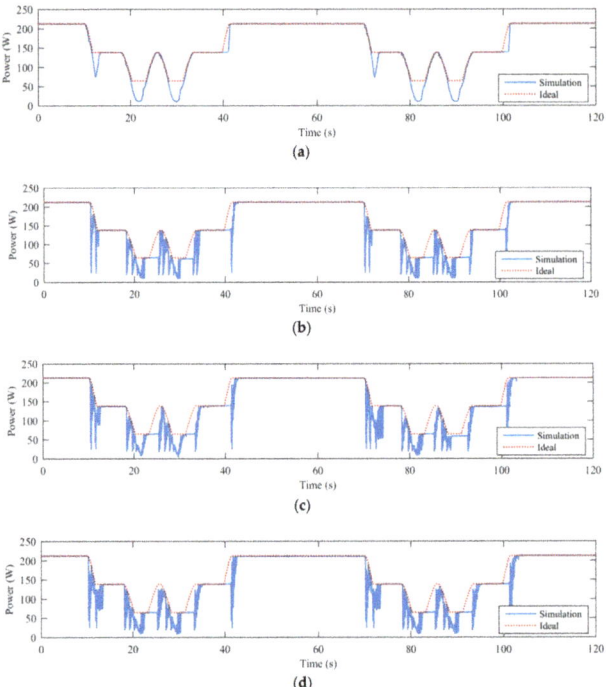

Figure 7. Rectangular shape—2 × 1: (**a**) P&O—B; (**b**) PSO 3 particles; (**c**) PSO 4 particles; (**d**) PSO 5 particles.

Table 3. Efficiency—Rectangular shapes (simulations and preliminary experimental tests).

MPPT	Efficiency					
	2 × 1		3 × 1		4 × 1	
	Sim.	Exp.	Sim.	Exp.	Sim.	Exp.
Perturb & Observe—A	95.34%	93.66%	96.19%	97.93%	95.37%	95.41%
Perturb & Observe—B	95.85%	95.54%	93.01%	93.29%	94.23%	92.76%
Perturb & Observe—C	93.40%	91.04%	94.34%	91.94%	93.48%	94.52%
Incremental Conductance	94.92%	92.74%	95.88%	94.97%	95.28%	95.68%
Constant Voltage	68.10%	69.37%	64.56%	65.38%	60.18%	60.09%
Open Voltage	79.33%	78.77%	73.56%	73.15%	71.38%	71.74%
Short Current Pulse	93.87%	93.17%	93.59%	94.25%	91.33%	89.30%
PSO—3 particles	92.54%	89.40%	93.48%	91.35%	88.77%	85.59%
PSO—4 particles	91.55%	91.65%	90.84%	88.02%	85.66%	82.86%
PSO—5 particles	90.80%	90.08%	91.76%	88.92%	89.42%	89.60%

4.2. Results: Trapezoidal-A Shape

Table 4 reports the efficiency calculated for all the dynamical shading scenarios characterized by Trapezoidal-A shape's objects. Figure 8 shows, as example, the diagrams of the harvested power (solid blue line) and maximum harvestable power (dotted red line) obtained in case of dynamical shading caused by the trapezoidal-A shape 3 × 1 shading object and controlling the off grid PV system, with the best performing traditional MPPT algorithm (P&O—A) and the best performing PSO-based MPPT algorithm (PSO—3p).

The obtained results are very similar to those obtained in the case of dynamical shading generated by rectangular objects. In this case the efficiency of the best PSO-based MPPT algorithm is again lower than the best hill climbing technique, namely 2.38% (2 × 1, P&O—B vs. PSO—3p) and 4.82% (3 × 1, P&O—A vs. PSO—3p) less respectively, and no specific rule determines the optimum number of particles.

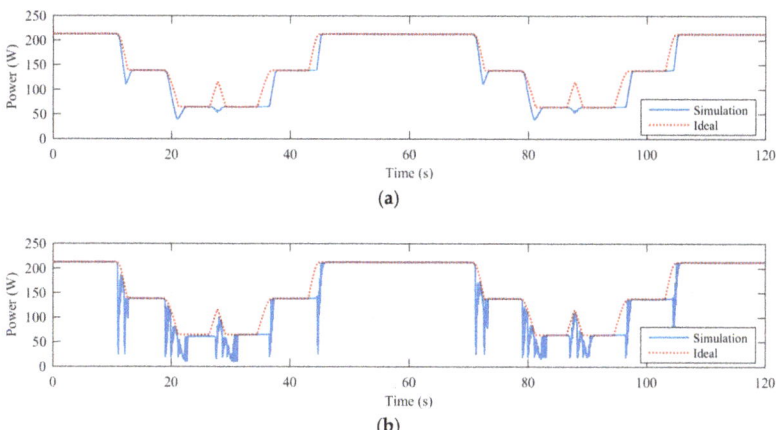

Figure 8. Trapezoidal-A shape—3 × 1: (**a**) P&O—A; (**b**) PSO 3 particles.

Table 4. Efficiency—Trapezoidal-A shapes (simulations).

MPPT	Efficiency		
	2 × 1	3 × 1	4 × 1
Perturb & Observe—A	95.37%	95.82%	95.95%
Perturb & Observe—B	97.59%	94.68%	95.30%
Perturb & Observe—C	93.28%	94.26%	93.69%
Incremental Conductance	96.03%	94.73%	94.72%
Constant Voltage	69.90%	66.41%	62.19%
Open Voltage	69.22%	73.29%	72.17%
Short Current Pulse	93.75%	93.47%	91.42%
PSO—3 particles	92.77%	93.44%	89.71%
PSO—4 particles	87.89%	90.91%	90.64%
PSO—5 particles	90.34%	91.57%	91.36%

4.3. Results: Trapezoidal-B Shape

Table 5 reports the efficiency calculated for all the dynamical shading scenarios characterized by Trapezoidal-B shape's objects. Figure 9 shows, as example, the diagrams of the harvested power (solid blue line) and maximum harvestable power (dotted red line) obtained in case of dynamical shading caused by the trapezoidal-A shape 4 × 1 shading object and controlling the off grid PV system, with the best performing traditional MPPT algorithm (P&O—A) and the best performing PSO-based MPPT algorithm (PSO—3p).

The obtained results are very similar to those obtained in the previous cases of dynamical shading. The efficiency of the best PSO-based MPPT algorithm ranges from a 3.42% (4 × 1, P&O—A vs. PSO—3p) to a 4.51% (2 × 1, P&O—B vs. PSO—3p) lower than the best hill climbing technique.

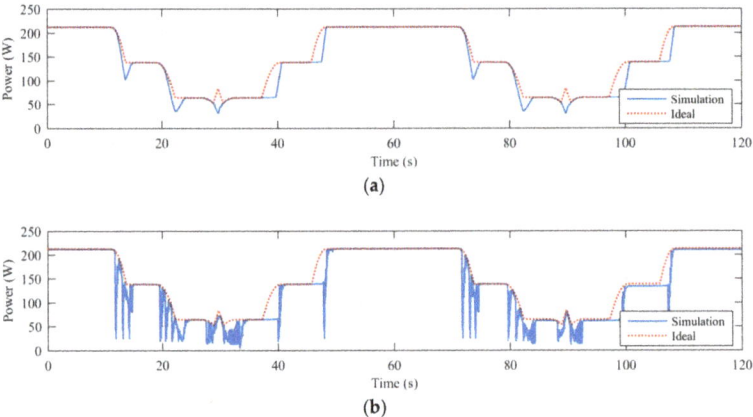

Figure 9. Trapezoidal-B shape—4 × 1: (**a**) P&O—A; (**b**) PSO 3 particles.

Table 5. Efficiency—Trapezoidal-B shapes (simulations).

MPPT	Efficiency		
	2 × 1	3 × 1	4 × 1
Perturb & Observe—A	97.39%	94.52%	95.41%
Perturb & Observe—B	97.91%	95.81%	91.14%
Perturb & Observe—C	95.62%	92.58%	93.47%
Incremental Conductance	97.06%	92.94%	92.83%
Constant Voltage	85.98%	69.60%	66.15%
Open Voltage	85.32%	68.92%	75.62%
Short Current Pulse	91.98%	91.77%	91.17%
PSO—3 particles	93.40%	90.47%	91.99%
PSO—4 particles	92.34%	91.82%	90.11%
PSO—5 particles	91.25%	89.27%	89.30%

4.4. Results: Trapezoidal-C Shape

Finally, Table 6 reports the efficiency calculated for all the dynamical shading scenarios characterized by the Trapezoidal-C shape's objects. Figure 10 shows, as example, the diagrams of the harvested power (solid blue line) and maximum harvestable power (dotted red line) obtained in case of dynamical shading caused by the trapezoidal-A shape 4 × 1 shading object and controlling the off grid PV system, with the best performing traditional MPPT algorithm (P&O—A) and the best performing PSO-based MPPT algorithm (PSO—5p).

The obtained results are again quite similar to those obtained in the previous cases. The efficiency of the best PSO-based MPPT algorithm in this last case ranges from 2.86% (3 × 1, P&O—A vs. PSO—3p) to 5.18% (2 × 1, P&O—B vs. PSO—5p) lower than the best hill climbing technique, confirming once more the same performance behavior, which in these simulations is independent of the particular applied dynamic shading conditions.

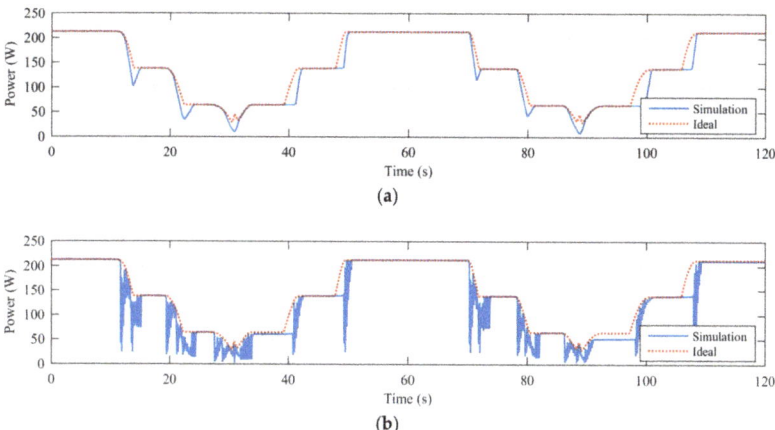

Figure 10. Trapezoidal-C shape—4 × 1: (**a**) P&O—A; (**b**) PSO 5 particles.

Table 6. Efficiency—Trapezoidal-C shapes (simulations).

MPPT	Efficiency		
	2 × 1	3 × 1	4 × 1
Perturb & Observe—A	95.06%	95.48%	95.64%
Perturb & Observe—B	97.65%	94.10%	95.22%
Perturb & Observe—C	93.35%	93.82%	93.19%
Incremental Conductance	95.34%	94.16%	94.31%
Constant Voltage	71.59%	67.37%	63.15%
Open Voltage	70.92%	73.81%	72.42%
Short Current Pulse	93.01%	92.81%	91.77%
PSO—3 particles	91.07%	92.62%	88.54%
PSO—4 particles	91.78%	92.02%	90.40%
PSO—5 particles	92.47%	91.61%	91.10%

5. Conclusions

Solar energy remains one of the best choices among renewables, being an inexhaustible energy source available everywhere with zero carbon emissions, but optimization in the performance of PV systems has become a priority.

This paper has proposed a comparison of traditional MPPT algorithms with an evolutionary based computational technique able to optimize extraction of solar power under dynamic environmental conditions. Partial shading conditions in different shape, size, and dynamics have been simulated, thus proving that dynamic conditions always represent a critical task for maximum power extraction. The effective use of a PSO-based algorithm has demonstrated that computational intelligence techniques can provide effective methods only if the population size increases, but under dynamic shading conditions computational time cannot be a negligible issue related to high conversion efficiency tasks. Moreover, this study has shown that PSO-based MPPT algorithm guarantees to effectively reach the global MPP, but the search of the global power point requires significant computational time, thus compromising conversion efficiency, especially in fast dynamic conditions.

Future works will include further improvements in the efficiency of the algorithm by reducing the computational time taken for scanning, choosing a more efficient approach that also considers hybrid strategies, and a deep experimental test campaign.

Author Contributions: Alberto Dolara has been the corresponding author and with Francesco Grimaccia contributed to the manuscript composition and test scenario developments. Alberto Dolara, Emanuele Ogliari and Marco Mussetta also implemented and ran the algorithms in the Matlab environment. Sonia Leva supervised the manuscript composition and the experimental validation, carried out at the Department of Energy, Politecnico di Milano (Italy).

Conflicts of Interest: The authors declare no conflict of interest.

Nomenclature

CI	Computational Intelligence
CV	Constant Voltage
DC/DC	Direct Current to Direct Current
GMPP	Global Maximum Power Point
IC	Incremental Conductance
MPP	Maximum Power Point
MPPT	Maximum Power Point Tracking
OV	Open Voltage
PSO	Particle Swarm Optimization
PV	Photovoltaic
P&O	Perturb and Observe
SC	Short Current Pulse

References

1. Hernandez, R.R.; Easter, S.B.; Murphy-Mariscal, M.L.; Maestre, F.T.; Tavassoli, M.; Allen, E.B.; Barrows, C.W.; Belnap, J.; Ochoa-Hueso, R.; Ravi, S.; et al. Environmental impacts of utility-scale solar energy. *Renew. Sustain. Energy Rev.* **2014**, *29*, 766–779. [CrossRef]
2. Turney, D.; Fthenakis, V. Environmental impacts from the installation and operation of large-scale solar power plants. *Renew. Sustain. Energy Rev.* **2011**, *15*, 3261–3270. [CrossRef]
3. Karavas, C.S.; Arvanitis, K.G.; Kyriakarakos, G.; Piromalis, D.D.; Papadakis, G. A novel autonomous PV powered desalination system based on a DC microgrid concept incorporating short-term energy storage. *Sol. Energy* **2018**, *159*, 947–961. [CrossRef]
4. Dolara, A.; Grimaccia, F.; Magistrati, G.; Marchegiani, G. Optimization Models for islanded micro-grids: A comparative analysis between linear programming and mixed integer programming. *Energies* **2017**, *10*, 1–20. [CrossRef]
5. Grimaccia, F.; Leva, S.; Mussetta, M.; Ogliari, E. ANN sizing procedure for the day-ahead output power forecast of a PV plant. *Appl. Sci.* **2017**, *7*, 622. [CrossRef]
6. Dolara, A.; Leva, S.; Ogliari, E.; Manzolini, G.; Ogliari, E. Investigation on performance decay on photovoltaic modules: Snail trails and cell microcracks. *IEEE J. Photovolt.* **2014**, *4*, 1204–1211. [CrossRef]
7. Dolara, A.; Lazaroiu, G.C.; Leva, S.; Manzolini, G.; Votta, L. Snail Trails and Cell Microcrack Impact on PV Module Maximum Power and Energy Production. *IEEE J. Photovolt.* **2016**, *6*, 1269–1276. [CrossRef]
8. Aghaei, M.; Grimaccia, F.; Gonano, C.A.; Leva, S. Innovative Automated Control System for PV Fields Inspection and Remote Control. *IEEE Trans. Ind. Electron.* **2015**, *62*, 7287–7296. [CrossRef]
9. Grimaccia, F.; Leva, S.; Niccolai, A. PV plant digital mapping for modules' defects detection by unmanned aerial vehicles. *IET Renew. Power Gener.* **2017**, *11*, 1221–1228. [CrossRef]
10. Siddique, N.; Adeli, H. *Computational Intelligence: Synergies of Fuzzy Logic, Neural Networks and Evolutionary Computing*; John Wiley & Sons: Chichester, UK, 2013.
11. Coello, C.A.; Lechuga, M.S. MOPSO: A proposal for multiple objective particle swarm optimization. In Proceedings of the Congress on Evolutionary Computation (CEC'02), Honolulu, HI, USA, 12–17 May 2002.
12. Berrera, M.; Dolara, A.; Faranda, R.; Leva, S. Experimental test of seven widely-adopted MPPT algorithms. In Proceedings of the 2009 IEEE Bucharest PowerTech: Innovative Ideas Toward the Electrical Grid of the Future, Bucarest, Romania, 28 June–2 July 2019.
13. Prasanth Ram, J.; Sudhakar Babu, T.; Rajasekar, N. A comprehensive review on solar PV maximum power point tracking techniques. *Renew. Sustain. Energy Rev.* **2017**, *67*, 826–847. [CrossRef]

14. Salas, V.; Olías, E.; Barrado, A.; Lázaro, A. Review of the maximum power point tracking algorithms for stand-alone photovoltaic systems. *Sol. Energy Mater. Sol. Cells* **2006**, *90*, 1555–1578. [CrossRef]
15. Karami, N.; Moubayed, N.; Outbib, R. General review and classification of different MPPT Techniques. *Renew. Sustain. Energy Rev.* **2017**, *68*, 1–18. [CrossRef]
16. Dolara, A.; Leva, S.; Magisrtrati, G.; Mussetta, M.; Ogliari, E. A novel MPPT algorithm for photovoltaic systems under dynamic partial shading-Recurrent scan and track method. In Proceedings of the 5th IEEE International Conference on Renewable Energy Research and Applications, Birmingham, UK, 20–23 November 2016.
17. Kollimalla, S.K.; Mishra, M.K. Variable Perturbation Size Adaptive P&O MPPT Algorithm for Sudden Changes in Irradiance. *IEEE Trans. Sustain. Energy* **2014**, *5*, 718–728.
18. Ananthi, C.; Kannapiran, B. Improved design of sliding-mode controller based on the incremental conductance MPPT algorithm for PV applications. In Proceedings of the 2017 IEEE International Conference on Electrical, Instrumentation and Communication Engineering, Karur, India, 27–28 April 2017.
19. Ma, J.; Man, K.L.; Ting, T.O.; Zhang, N.; Lei, C.; Wong, N. A Hybrid MPPT Method for Photovoltaic Systems via Estimation and Revision Method. In Proceedings of the IEEE International Symposium on Circuits and Systems (ISCAS), Beijing, China, 19–23 May 2013.
20. Sher, H.A.; Addoweesh, K.E.; Haddad, K.A. An Efficient and Cost-Effective Hybrid MPPT Method for a Photovoltaic Flyback Micro-Inverter. *IEEE Trans. Sustain. Energy* **2017**, *PP*, 1. [CrossRef]
21. Hanafiah, S.; Ayad, A.; Hehn, A.; Kennel, R. A hybrid MPPT for quasi-Z-source inverters in PV applications under partial shading condition. In Proceedings of the 11th IEEE International Conference on Compatibility, Power Electronics and Power Engineering, Cadiz, Spain, 4–6 April 2017.
22. Ishaque, K.; Salam, Z. A Deterministic Particle Swarm Optimization Maximum Power Point Tracker for Photovoltaic System under Partial Shading Condition. *IEEE Trans. Ind. Electron.* **2013**, *60*, 3195–3206. [CrossRef]
23. Hohm, D.P.; Ropp, M.E. Comparative Study of Maximum Power Point Tracking Algorithms Using an Experimental, Programmable, Maximum Power Point Tracking Test Bed. In Proceedings of the IEEE Photovoltaic Specialist Conference, Anchorage, AK, USA, 15–22 September 2000.
24. Aghaei, M.; Dolara, A.; Grimaccia, F.; Leva, S.; Kania, D.; Borkowski, J. Experimental Comparison of MPPT Methods for PV Systems Under Dynamic Partial Shading Conditions. In Proceedings of the 16th International Conference on Environment and Electrical Engineering, Florence, Italy, 7–10 June 2016.
25. Dolara, A.; Lazaroiu, G.C.; Leva, S.; Manzolini, G. Experimental investigation of partial shading scenarios on PV (photovoltaic) modules. *Energy* **2013**, *55*, 466–475. [CrossRef]
26. Dolara, A.; Leva, S.; Manzolini, G. Comparison of different physical models for PV power output prediction. *Sol. Energy* **2015**, *119*, 83–99. [CrossRef]
27. Dolara, A.; Lazaroiu, G.C.; Ogliari, E. Efficiency analysis of PV power plants shaded by MV overhead lines. *Int. J. Energy Environ. Eng.* **2016**, *7*, 115–123. [CrossRef]

 © 2018 by the authors. Licensee MDPI, Basel, Switzerland. This article is an open access article distributed under the terms and conditions of the Creative Commons Attribution (CC BY) license (http://creativecommons.org/licenses/by/4.0/).

Article

Metaheuristic Algorithm for Photovoltaic Parameters: Comparative Study and Prediction with a Firefly Algorithm

Mohamed Louzazni [1,*], Ahmed Khouya [1], Khalid Amechnoue [1], Alessandro Gandelli [2], Marco Mussetta [2] and Aurelian Crăciunescu [3]

1. Mathematics Informatic & Applications Team, National School of Applied Sciences, Abdelmalek Essaadi University, Tanger 1818, Morocco; ahmedkhouya3@yahoo.fr (A.K.); kamechnoue@gmail.com (K.A.)
2. Department of Energy, Politecnico di Milano, 20156 Milano, Italy; alessandro.gandelli@polimi.it (A.G.); marco.mussetta@polimi.it (M.M.)
3. Electrical Engineering Department, University Politehnica of Bucharest, Bucharest 060042, Romania; aurelian.craciunescu@upb.ro
* Correspondence: louzazni@msn.com; Tel.: +212-6138-81279

Received: 31 December 2017; Accepted: 20 February 2018; Published: 27 February 2018

Featured Application: The parameter identification of solar cell and photovoltaic module are used for evaluation, control and optimization of photovoltaic systems.

Abstract: In this paper, a Firefly algorithm is proposed for identification and comparative study of five, seven and eight parameters of a single and double diode solar cell and photovoltaic module under different solar irradiation and temperature. Further, a metaheuristic algorithm is proposed in order to predict the electrical parameters of three different solar cell technologies. The first is a commercial RTC mono-crystalline silicon solar cell with single and double diodes at 33 °C and 1000 W/m^2. The second, is a flexible hydrogenated amorphous silicon a-Si:H solar cell single diode. The third is a commercial photovoltaic module (Photowatt-PWP 201) in which 36 polycrystalline silicon cells are connected in series, single diode, at 25 °C and 1000 W/m^2 from experimental current-voltage. The proposed constrained objective function is adapted to minimize the absolute errors between experimental and predicted values of voltage and current in two zones. Finally, for performance validation, the parameters obtained through the Firefly algorithm are compared with recent research papers reporting metaheuristic optimization algorithms and analytical methods. The presented results confirm the validity and reliability of the Firefly algorithm in extracting the optimal parameters of the photovoltaic solar cell.

Keywords: solar cell; metaheuristic algorithm; electrical parameters; analytical methods; firefly algorithm; statistical errors

1. Introduction

The use of renewable energy sources is rapidly developing, and the application of solar energy focusing on photovoltaic systems is becoming increasingly popular [1,2]. The major challenge in photovoltaics system is posed by the instability, nonlinearity and complexity of the current-voltage and power-voltage characteristics equation. The relation between photovoltaic current and voltage is both implicit and nonlinear [3–6] and it depends on several factors such as module temperature, solar radiation and its distribution, spectrum, cable losses, dust accumulation, shading and soiling [7,8]. Therefore, it is vital to produce a more accurate mathematical model that can better reveal the actual behavior and represent the relationship between current and voltage. In this context, many mathematical models have been developed in the literature to describe the electric, dynamic and thermal behavior of photovoltaic cell/module with a different level of complexity. In particular, the solar cell can be modelled as a static model for DC/DC (direct current), or as a dynamic model for DC/AC (alternating current) with capacitance and parallel dynamic resistance, with diode and photocurrent as proposed in [9–11]. They can be classified globally into two categories: implicit and explicit models [12,13]. The former [5,14–16] need iterative numerical methods to solve the nonlinear current-voltage equation. On the other hand, the latter models are based on simple analytical expressions [4,17–22]. Different physical models were compared on photovoltaic power output prediction in [23] and available models of solar cell are presented in [24]. A different photovoltaic model used for 24-hour-ahead forecasting using neural network is presented in [25], while a comparison between physical and hybrid methods is given in [26] and artificial neural network models are employed in [27]. These models differ mainly by the number of diodes, the presence or absence of a shunt resistor, and by the numerical methods used to determine the unknown parameters. Further, the two diodes model is known as the most accurate model for representing the equivalent electrical circuit. While the single diode model is the most commonly used of the two types; in the simplified four-parameter model neglecting shunt resistance by assuming it as infinite value, and in five-parameter models by maintaining the effect of the shunt resistance. The five and seven parameters models evaluate the photocurrent, the saturation current, the series and shunt resistors and the quality factor of the diode. The eight parameters model adds build-in voltage, thickness, average mobility-lifetime.

The exponential non-linearity of current-voltage equations causes many difficulties in prediction and extraction of the electric, dynamic or thermal parameters [28] while, the implicit models are not capable of determining the behavior of the photovoltaic cell/module under many effects. Furthermore, solar cell models have multi-modal objective functions and model parameters vary with operational conditions such as temperature and irradiance. The main problem is to identify the optimal parameter values such as photo-generated current, diode saturation current, series resistance, and diode quality factor. Over the years, various papers have been presented and developed different techniques to identify the optimal values of the electric parameters to describe the behavior of the characteristics. These can be categorized into analytical methods, numerical methods and metaheuristic methods. There are several analytical and numerical (generally gradient-based) methods, as described in Table 1.

Table 1. A list of analytical and numerical methods employed in the literature.

Optimization Method	Reference
Least squares and Newton-Raphson method	[29]
Iterative curve fitting	[30]
Lambert W-functions	[20,31–35]
Integral-based linear least square identification method	[36,37]
Linear interpolation/extrapolation	[38]
Chebyshev polynomials	[39]
Taylor's series expansion	[40]
Padé approximants	[41]
Symbolic function	[42]
Analytical mathematical method	[43–45]
Simple methods based on measured points	[46]

Metaheuristic methods are powerful in local searches, but they tend to get trapped in locally optimal values and depend on the photovoltaic module's manufacturer's data such as open circuit, short circuit, and maximum power points. Since the photovoltaic cell has triple non-linearity in current-voltage, power-voltage and in intrinsic parameters, deterministic methods cannot extract parameters accurately based on current, voltage and current derivatives with respect to the voltage at short circuit current, maximum power and open circuit voltage. The derivation imposes several model restrictions such as convexity, continuity and differentiability conditions; moreover, the approximations also reduce accuracy. Due to their great potential in modern global optimization resolution for nonlinear and complex systems, the use of metaheuristic bioinspired optimization algorithms to carry out minimizing procedures has received considerable attention. Metaheuristic methods are stochastic methods inspired by various natural phenomenon, as listed in Table 2. They have been proven to be a promising alternative to deterministic methods applied to the parameter identification of solar/photovoltaic models.

Table 2. A list of metaheuristic methods employed in the literature.

Metaheuristic Methods	Reference
Levenberg-Marquardt algorithm combined with Simulated Annealing	[47]
Artificial Bee Swarm	[48]
Artificial Bee Colony	[49]
Hybrid Nelder-Mead and Modified Particle Swarm	[50]
Firefly Algorithm	[51–53]
Self-Organizing Migrating Algorithm	[54]
Pattern Search	[55]
Genetic Algorithm	[56,57]
Simulated Annealing algorithm	[58]
Repaired Adaptive Differential Evolution	[59]
Particle Swarm Optimization	[60]
Bird Mating Optimization approach	[61]

However, the cited algorithms are usually trapped at local optima and they have large error values [62]. In fact, the performance of these algorithms highly depends on the settings of specific parameters, such as, for instance, the mutation probability, crossover probability, and the selection operator in the genetic algorithm, as well as the inertia weight, and social and cognitive parameters in particle swarm optimization. Therefore, researchers are still searching for powerful algorithms capable of predicting the optimal parameters of different technology under various conditions with less errors.

Metaheuristic bioinspired algorithms have been suggested for parameter extraction and have become an important part of modern optimization. Most metaheuristic algorithms are based on natural or artificial swarm intelligence. Particle swarm optimization is a good example, it mimics the swarming behavior of bees and birds [62]. Recently, a new metaheuristic search algorithm called the firefly algorithm (FA) has been proposed and developed by X. Yang [63]. The FA is a nature-inspired stochastic optimization algorithm based on the flashing patterns and behavior of swarming fireflies [64]. The FA has become an increasingly valuable tool of swarm intelligence that has been applied in almost all areas of optimization, as well as in engineering practice [65]. It uses a kind of randomization by searching for set solutions, inspired by the flashing lights of fireflies in nature. This algorithm differs from many swarm intelligence techniques [65] for these two features:

- the first is the so-called local attraction, since the light intensity decreases with distance (the attractions of fireflies can be local or global and depend on the absorbing coefficient);
- the second is related to the subdivision of fireflies and their regrouping into subgroups because a neighboring attraction is stronger than a long-distance attraction, and each subgroup will swarm around a local mode, making the firefly algorithm suitable for multimodal global optimization problems [66].

In [67] the authors provide a detailed background and analysis of the firefly algorithm and test it in a wide range of problems to solve multi-objective dispatch problems.

In this paper, the authors propose a comparison among bioinspired algorithms for the prediction of solar cell and photovoltaic module parameters. The goal is to minimize the multi-objective functions adapted to minimize the absolute errors between experimental and calculated current-voltage data under inequality constraint functions. Three different cases are examined as follows: single and double diode models of a commercial mono-crystalline silicon solar cell (R.T.C France company) at 33 °C, based on current-voltage experimental data recorded in [29]; (ii) flexible dual junction amorphous hydrogenated silicon a-Si:H solar cell under standard sunlight, based on data obtained in a light intensity of 1000 W/m² and at a temperature of 300 K [54]; (iii) a Photowatt-PWP 201 photovoltaic module which 36 polycrystalline silicon cells are connected in series and the data is measured at an irradiance of 1000 W/m², and a temperature of 25 °C [29]. To verify the performance of the proposed approach and the quality of the obtained results, statistical analyses are carried out to measure the accuracy of the calculated parameters and model suitability. The results obtained are compared with recent techniques such as the Biogeography-Based Optimization algorithm with Mutation strategies (BBO-M) [68], Levenberg-Marquardt algorithm combined with Simulated Annealing (LMSA) [47], Artificial Bee Swarm Optimization algorithm [48], Artificial Bee Colony optimization (ABC) [49], hybrid Nelder-Mead and Modified Particle Swarm Optimization (NM-MPSO) [50], Repaired Adaptive Differential Evolution (RADE) [59], Chaotic Asexual Reproduction Optimization (CARO) [69] for solar cell single and double diodes. For organic flexible hydrogenated amorphous silicon, a-Si:H solar cell will be compared with the Quasi-Newton (Q-N) method and Self-Organizing Migrating Algorithm (SOMA) [54]. The optimal parameters of Photowatt-PWP 201 are compared with the Newton-Raphson [29] Pattern Search (PS) [55], Genetic algorithm (GA) [56] and Simulated Annealing algorithm (SA) [58]. The obtained results are in accordance with experimental data, there is good agreement for most of the extracted parameters and the proposed algorithm outperformed the compared techniques.

2. Presentation and Modelling of the Solar Cell

The electrical behavior of the solar cell is modelled by its outputs current versus voltage characteristic. Further, a solar cell is mathematically modelled in two common methods [24,70], single diode (SDM) and double diode (DDM), with consider parasitic phenomena by series and shunt resistances. Moreover, the flexible hydrogenated amorphous silicon a-Si:H solar cell with loss current I_{rec} is paralleled with the original photo-generated current source and the current sink representing the recombination current in the *i*-layer of a P-I-N solar cell [71–75]. The two models are given in Figure 1.

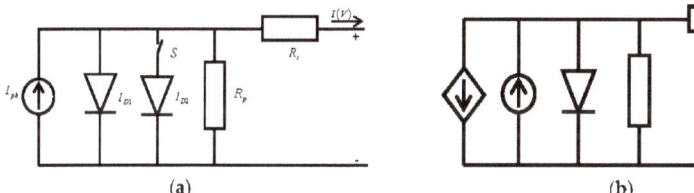

Figure 1. Equivalent circuit solar cell model: (a) single and double diode, (b) flexible hydrogenated amorphous silicon a-Si:H.

The current-voltage behavior of a solar cell is described according to the electrical intrinsic parameters and nonlinear implicit equation, for a given illumination and temperature.

$$I = I_{ph} - I_{SD1}\left[\exp\left(\frac{V + IR_s}{a_1 V_T}\right) - 1\right] - SI_{SD2}\left[\exp\left(\frac{V + IR_s}{a_2 V_T}\right) - 1\right] - \frac{V + IR_s}{R_P} \quad (1)$$

where, I_{ph} is the photocurrent, I_{SD1} and I_{SD2} are the saturation currents, a_1 and a_2 are the diffusion and recombination diode quality factors; R_s and R_p are the resistances in series and parallel, respectively, V_T is the thermal voltage (which will be defined in the followings), and:

$$S = \begin{cases} SDM & \text{for } Open = 0 \\ SDDM & \text{for } Close = 1 \end{cases} \quad (2)$$

The current-voltage characteristic of a flexible solar cell is:

$$I = I_{ph}\left(1 - \frac{d_i^2}{(\mu\tau)_{eff}[V_{bi} - (V + IR_s)]}\right) - I_s\left[\exp\left(\frac{(V + IR_s)}{aV_T}\right) - 1\right] - \frac{V + IR_s}{R_P} \quad (3)$$

where, the voltage V_{bi} represents the built-in field voltage over the i-layer, in single junction amorphous silicon solar cells, and in [76] it has been determined to be in the range 0.9 V; d_i represents the thickness of the i-layer, the effective $\mu\tau$-product $(\mu\tau)_{eff}$ represents average mobility-lifetime product for election and hole, and quantifies the quality of the active layer in terms of recombination of photo-generated carriers. The thermal voltage is $V_T = KT/q$ where K is Boltzmann's constant, T is the cell absolute temperature in Kelvin and q is the electronic charge, a is the diode quality factor.

The photocurrent I_{ph} describes the irradiation dependent recombination in i-layer and reduced by the recombination current, as follows:

$$I_{rec} = I_{ph}\left[\left(\frac{\mu\tau}{d_i^2}\right)(V_{bi} - (V + IR_s))\right]^{-1} \quad (4)$$

where, I_{rec} is the current sink and it represents the recombination current in the i-layer of a P-I-N; the current through the diode represents the diffusion process of charge carriers and the last term represents the shunt leakage current I_p and is modelled as a space charge limited current [77,78].

In Equations (1) and (3), the five, seven and eight parameters which define the current versus voltage relation of solar cell and photovoltaic module, vary in accordance with solar irradiance, cell temperature and depend on reference values reported on datasheet.

3. Problem Formulation

The solar cell can be modelled by using the single diode model, double diode or multi-diode models. The objective function is defined from Equations (1) and (3), several research papers use different functions, for example, [48–50,59,68,69] use the root mean square error (RMSE), [47] use the sum of squared error (SSE). In [55,58] the individual absolute error (IAE) is used and [79] use the mean absolute errors (MAE). However, the objective function was used to minimize the vertical distance between the experimental points and the theoretical curve. In this paper, we use separate fitting for different regions in the current-voltage characteristics (Figure 2), because the current error is more important for small voltages due to the strongly varying slope of the curve, while the voltage error is more important for large voltages approaching an open circuit.

During the optimization process, each i-th solution is defined by a vector X_i, where X is a candidate set of parameters defined as follows:

- for a single diode: $X = x_1 = \begin{bmatrix} I_{ph} & I_{SD} & a & R_S & R_P \end{bmatrix}$;
- for a double diode: $X = x_2 = \begin{bmatrix} I_{ph} & I_{SD1} & I_{SD2} & a_1 & a_2 & R_S & R_P \end{bmatrix}$;
- for a flexible solar cell: $X = x_3 = \begin{bmatrix} I_{ph} & d_i & \mu\tau & V_{bi} & R_s & I_0 & a & R_{sh} \end{bmatrix}$.

The objective functions must be minimized with respect to the limits of parameters x_1, x_2 and x_3. The Equations (1) and (3) is rewritten in the following homogeneous equations.

For a single and double diode:

$$F_{1/2}(V, I) = I - I_{ph} + I_{SD1}\left[\exp\left(\frac{V + IR_s}{a_1 V_T}\right) - 1\right] + SI_{SD2}\left[\exp\left(\frac{V + IR_s}{a_2 V_T}\right) - 1\right] + \frac{V + IR_s}{R_P} \quad (5)$$

For flexible hydrogenated amorphous silicon, a-Si:H:

$$G(V, I, x_3) = I - I_{ph}\left(1 - \frac{d_i^2}{(\mu\tau)_{eff}[V_{bi} - (V + IR_s)]}\right) + I_s\left[\exp\left(\frac{(V + IR_s)}{aV_T}\right) - 1\right] + \frac{V + IR_s}{R_P} \quad (6)$$

The cost function of current error ε_1 near the short circuit (zone 1) is:

$$\varepsilon_1 = \sqrt{\frac{1}{m_1} \sum_{Zone1} (I_i(V_i, I_i, x) - I_{PV\_exp\,-i})^2} \qquad (7)$$

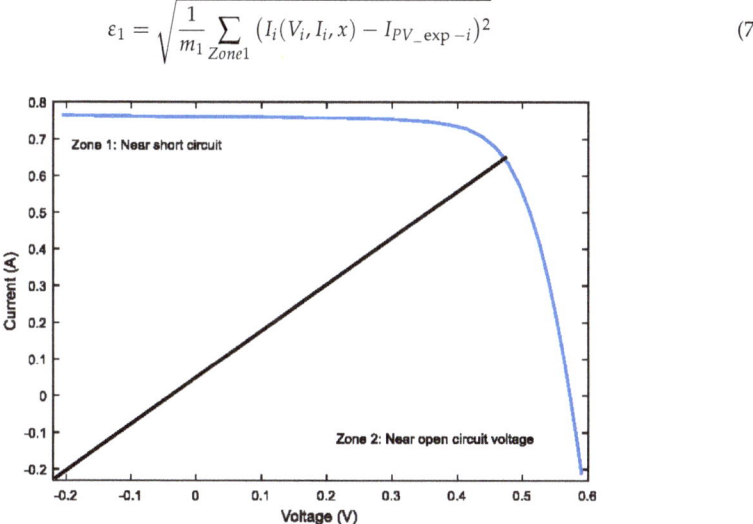

Figure 2. The characteristic current-voltage division in two zones.

The cost function of voltage error ε_2 is the horizontal distance between the experimental point and calculated curve; it is defined near the open circuit (zone 2) as:

$$\varepsilon_2 = \sqrt{\frac{1}{m_2} \sum_{Zone2} (V_i(V_i, I_i, x) - V_{PV\_exp\,-i})^2} \qquad (8)$$

Where m is the number of experimental data, V_i and I_i are the i-th simulated and experimental current and voltage value, respectively. The overall objective function, i.e., the global error ε in the two zones, is the sum of current and voltage errors and is defined as:

$$f(X) = \varepsilon = \varepsilon_1 + \varepsilon_2 \qquad (9)$$

The objective function constraints for each model are presented in the following equations. For a single and double diode:

$$\min \varepsilon_{SD}$$
$$\text{subject to} \begin{cases} 0 \leq I_{ph} \leq 1 \\ 0 \leq I_{SD} \leq 1 \times 10^{-7} \\ 1 \leq a \leq 2 \\ 0 \leq R_S \leq 0.5 \\ 0 \leq R_P \leq 100 \end{cases}$$

$$\min \varepsilon_{SD}$$
$$\text{subject to} \begin{cases} 0 \leq I_{ph} \leq 1 \\ 0 \leq I_{SD1} \leq 1 \times 10^{-7} \\ 0 \leq I_{SD2} \leq 1 \times 10^{-7} \\ 1 \leq a_1 \leq 2 \\ 1 \leq a_2 \leq 2 \\ 0 \leq R_S \leq 0.5 \\ 0 \leq R_P \leq 100 \end{cases}$$

While, for flexible hydrogenated amorphous silicon, a-Si:H:

$$\min \varepsilon_F$$

$$\text{subject to} \begin{cases} 0 \leq I_{ph} \leq 0.8\,\text{A} \\ 0 \leq d_i \leq 4\,\text{nm} \\ 0 \leq \mu\tau \leq 5 \times 10^{-5}\,\text{cm}^2/\text{V} \\ 0 \leq V_{bi} \leq 1\,\text{V} \\ 0 \leq I_S \leq 4 \times 10^{-18}\,\text{A} \\ 1 \leq a \leq 1 \\ 0 \leq R_S \leq 0.5\,\Omega \\ 0 \leq R_P \leq 20\,\Omega \end{cases}$$

4. Firefly Optimization Algorithm

The Firefly algorithm is a swarm intelligence algorithm for optimization problems. It was introduced in 2009 at Cambridge University by Yang [64], and it is inspired by the flashing patterns and behavior of tropical fireflies at night, and it is flexible and easy to implement. The Firefly algorithm is a bio-inspired metaheuristic algorithm and a random optimization, which is capable of converging to a global solution of an optimization problem. It uses the following three idealized rules [63–67]:

1. No sex distinctions, i.e., fireflies are attracted to other fireflies regardless of their sex.
2. The degree of the attractiveness of a firefly is proportional to its brightness, thus for any two flashing fireflies, the less bright one will move towards the brighter one; the more brightness, the less the distance between two fireflies. If there is no brighter firefly, it will move randomly.
3. The brightness of a firefly is determined by the value of the objective function.

The basic rules of this algorithm were designed to primarily solve continuous problems. To design the Firefly algorithm properly, two critical issues need to be defined: the attractiveness and the variation of the light intensity.

4.1. Attractiveness

In the Firefly algorithm, the variation of the light intensity and the formulation of the attractiveness play a vital role. The intensity of light or brightness $I(r_{ij})$ is inversely proportional to the square of the distance r_{ij} [64,66] and the relative brightness of each firefly is expressed in the following Gaussian form:

$$I(r_{ij}) = I_0 e^{-\lambda r_{ij}^2} \tag{10}$$

where, $I(r_{ij})$ is the light intensity at a distance r_{ij}, I_0 is the maximum brightness (the absolute brightness at the source point $r_{ij} = 0$) which is related to the objective function value. The higher value of the objective function is the higher I_0 is and λ is the light absorption coefficient, which is set to reflect that brightness increases gradually with the increase in distance and the absorption of the medium r_{ij} is the Euclidean distance between firefly i and firefly j. The attractiveness of each firefly [56] is expressed in the form

$$\beta(r_{ij}) = \beta_0 e^{-\lambda r_{ij}^2} \tag{11}$$

where, β_0 is the maximum attractiveness (the attractiveness at $r_{ij} = 0$, the largest value of the firefly to attract another, is typically set to 1). However, computationally, computing $1/\left(1 + \lambda r_{ij}^2\right)$ is easier than $e^{-\lambda r_{ij}^2}$ [64] and the intensity can be written as:

$$I(r_{ij}) = \frac{I_0}{1 + \lambda r_{ij}^2} \tag{12}$$

Similarly, the attractiveness of a firefly can be approximated as follows:

$$\beta(r_{ij}) = \frac{\beta_0}{1 + \lambda r_{ij}^2} \qquad (13)$$

4.2. Distance and movement

We suppose a firefly located at $x_i = (x_1^i, x_2^i \ldots x_k^i)$ is brighter than another firefly located at $x_j = (x_1^j, x_2^j \ldots x_k^j)$, the firefly located at x_i will move towards x_j. The distance between any two fireflies i and j at x_i and x_j is the Euclidean distance given by [64,66] as follows:

$$r_{ij} = |x_i - x_j| = \sqrt{\sum_k^d \left(x_{i,k} - x_{j,k}\right)^2} \qquad (14)$$

where, d is the dimension, $x_{i,k}$ is the k-th component of the spatial coordinate x_i of i-th firefly the movement of a firefly i is attracted to another more attractive firefly j and the update location is determined by

$$x_{i+1} = x_i + \beta_0 e^{-\lambda r_{ij}^2}(x_j - x_i) + \alpha\left(rand - \frac{1}{2}\right) \qquad (15)$$

The first term is the current position of a firefly [66], the second term is used for considering a firefly's attractiveness to light intensity seen by adjacent fireflies and the third term is used for the random movement of a firefly in case there are not any brighter ones. The coefficient α is a randomization parameter determined by the problem of interest, while rand is a random-number drawn from a Gaussian distribution or uniform distribution at time t, if $\beta_0 = 0$, it becomes a simple random walk. In the implementation of the algorithm we will use $\beta_0 = 0$, $\alpha = 0.25$ and the attractiveness or absorption coefficient $\lambda = 1$ which guarantees a quick convergence of the algorithm to the optimal solution. The concept of the firefly-based algorithm is presented in Figure 3. Moreover, Figure 4 shows the here considered implementation of FA for the specific problem and cost function given in Equation (9), as defined in Section 3.

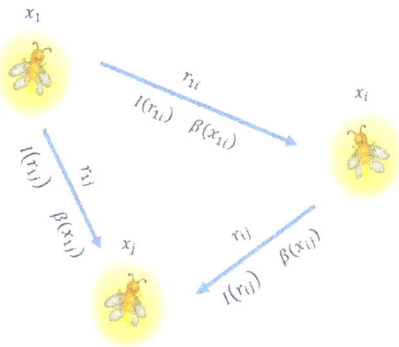

Figure 3. A conceptual view of the firefly algorithm relationships, including locations x, distance r, brightness $I(r)$, and attractiveness $\beta(r)$.

```
Init
Random initial population of N fireflies in locations $X_i$ ($i = 1 ... N$)
Definition of initial light intensity $I_i = f(X_i)$    % $f(X_i)$ = cost function from Equation (9)
Definition of light absorption coefficient $\lambda$
while $k < K$         % $K$ = maximun number of iterations
    for $i = 1:N$
        for $j = 1:N$
            if $I_j > I_i$
                Update attractiveness $\beta_{ij} = \beta_0 e^{-\lambda r_{ij}}$
                Update position $X_i$ by moving firefly $i$ towards $j$ in $d$-dimension domain
                Update light intensity $I_i$ by means of cost function evaluation: $I_i = f(X_i)$
            end
        end
    end
    Rank the N fireflies and find the current best $X_{best}^{(k)}$
end
```

Figure 4. A pseudocode for the considered firefly algorithm implementation.

5. Results, Discussions and Comparison

In order to evaluate the efficiency of the Firefly algorithm in the estimation of the solar cell, the photovoltaic module and the array parameter extraction, the results are compared with analytical methods, numerical methods and metaheuristic algorithm to validate the effectiveness of the algorithm. In order, to compare it with other algorithms, a benchmark commercial solar cell and benchmark photovoltaic module are selected in single diode, double diode and photovoltaic module models are considered. The study test cases are designated as follows:

- Test scenario 1: Apply to commercial solar cell for both single diode and double model under standard irradiance level with relevant example comparisons to other methods.
- Test scenario 2: Apply to a flexible hydrogenated amorphous silicon a-Si:H photovoltaic cell using single diode module.
- Test scenario 3: Apply to a commercial photovoltaic array using the single diode model, with 36 solar cells connected in series.

The current-voltage measurements are collected from [29,54] and have been widely used by different papers to test electric circuit models, modelling or translate the current versus voltage and technique for parameter extraction. Furthermore, statistical analyses are carried out to measure the accuracy of the estimated parameters and model suitability.

5.1. Case 1: Single and Double Diode Model (RTC France Company)

The proposed algorithm is applied first to extract the electrical intrinsic parameters values for single and double diode models of a 57-mm-diameter commercial (RTC France) silicon solar cell under 1000 W/m^2 at 33 °C. The extracted parameters are compared with those found by: Biogeography-Based Optimization algorithm with Mutation strategies (BBO-M) [68], Levenberg-Marquardt algorithm combined with Simulated Annealing (LMSA) [47], Artificial Bee Swarm Optimization algorithm [48], Artificial Bee Colony optimization (ABC) [49], hybrid Nelder-Mead and Modified Particle Swarm Optimization (NM-MPSO) [50], Repaired Adaptive Differential Evolution (RADE) [59], Chaotic

Asexual Reproduction Optimization (CARO) [69], and the results for each model are reported in Tables 3 and 4.

Table 3. Comparison of various parameter identification techniques for single diode model (RTC France Company). FA: Firefly Algorithm; BBO-M: Biogeography-Based Optimization with Mutation strategies; RADE: Repaired Adaptive Differential Evolution; LMSA: Levenberg-Marquardt algorithm combined with Simulated Annealing; CARO: Chaotic Asexual Reproduction Optimization; ABC: Artificial Bee Colony optimization; NM-MPSO: hybrid Nelder-Mead and Modified Particle Swarm Optimization.

Approaches	Parameter				
	I_{ph} (A)	I_0 (μA)	a	R_s (Ω)	R_p (Ω)
FA	0.76069712	0.4324411	1.45245666	0.03341059	53.40180803
BBO-M	0.76078	0.31874	1.47984	0.03642	53.36227
RADE	0.760776	0.323021	1.481184	0.036377	53.718526
LMSA	0.76078	0.31849	1.47976	0.03643	53.32644
CARO	0.76079	0.31724	1.48168	0.03644	53.0893
ABC	0.7608	0.3251	1.4817	0.0364	53.6433
NM-MPSO	0.76078	0.32306	1.48120	0.03638	53.7222

Table 4. Comparison of various parameter identification techniques for a double diode model (RTC France Company).

Approaches	Parameter						
	I_{ph} (A)	I_{01} (μA)	I_{02} (μA)	a_1	a_2	R_s (Ω)	R_p (Ω)
FA	0.760820	0.591126	0.245384	1.0246	1.3644	0.036639	55.049
RADE	0.760781	0.225974	0.749347	1.451017	2.0000	0.036740	55.485443
CARO	0.76075	0.29315	0.09098	1.47338	1.77321	0.03641	54.3967
ABSO	0.76078	0.26713	0.38191	1.46512	1.98152	0.03657	54.6219
ABC	0.7608	0.0407	0.2874	1.4495	1.4885	0.0364	53.7804
NM-MPSO	0.76078	0.22476	0.75524	1.45054	1.99998	0.03675	55.5296

To confirm the accuracy of the extracted optimal values found by the Firefly algorithm, the calculated currents for the single and double diode model by optimized parameters are summarized in Tables 5 and 6 compared with individual absolute error (IAE).

$$IAE = |I_{measured} - I_{estimated}| \quad (16)$$

Table 5. Calculated current and compared IAE for single diode (RTC France Company).

Item	V_{Exp} (V)	I_{Exp} (A)	$I_{Calculated}$ (A)	FA (A)	Individual Absolute Error (IAE)		
					RADE	BBO-M	NM-MPSO
1	−0.2057	0.7640	0.76407143	7.1420×10^{-5}	9.5590×10^{-5}	6.0000×10^{-6}	8.7000×10^{-5}
2	−0.1291	0.7620	0.76263790	6.3789×10^{-4}	6.6611×10^{-4}	6.0400×10^{-4}	6.6200×10^{-4}
3	−0.0588	0.7605	0.76132213	8.2213×10^{-4}	8.5473×10^{-4}	8.1700×10^{-4}	8.5400×10^{-4}
4	0.0057	0.7605	0.76015347	3.4652×10^{-4}	3.5034×10^{-4}	3.6400×10^{-4}	3.4600×10^{-4}
5	0.0646	0.7600	0.75905434	9.4565×10^{-4}	9.4298×10^{-4}	9.4600×10^{-4}	9.4500×10^{-4}
6	0.1185	0.7590	0.75804099	9.5900×10^{-4}	9.5528×10^{-4}	9.4300×10^{-4}	9.5700×10^{-4}
7	0.1678	0.7570	0.75702642	2.6419×10^{-5}	9.5100×10^{-5}	1.2000×10^{-4}	9.1000×10^{-5}
8	0.2132	0.7570	0.75614154	8.5846×10^{-4}	8.4950×10^{-4}	8.1700×10^{-4}	8.5800×10^{-4}
9	0.2545	0.7555	0.75509107	4.0892×10^{-4}	4.1823×10^{-4}	3.6100×10^{-4}	4.1300×10^{-4}
10	0.2924	0.7540	0.75367808	3.2191×10^{-4}	3.2967×10^{-4}	2.7600×10^{-4}	3.3600×10^{-4}
11	0.3269	0.7505	0.75111180	6.1180×10^{-4}	8.9542×10^{-4}	9.5300×10^{-4}	8.8800×10^{-4}
12	0.3585	0.7465	0.74691657	4.1656×10^{-4}	8.5737×10^{-4}	9.1400×10^{-4}	8.4800×10^{-4}
13	0.3873	0.7385	0.73945849	9.5848×10^{-4}	1.6042×10^{-3}	1.6680×10^{-3}	1.5960×10^{-3}
14	0.4137	0.7280	0.72757692	4.2308×10^{-4}	5.9912×10^{-4}	5.8300×10^{-4}	6.0400×10^{-4}

Table 5. Cont.

Item	V_{Exp} (V)	I_{Exp} (A)	$I_{Calculated}$ (A)	FA (A)	Individual Absolute Error (IAE)		
					RADE	BBO-M	NM-MPSO
15	0.4373	0.7065	0.70650197	1.9700×10^{-6}	4.4631×10^{-4}	4.8500×10^{-4}	4.5200×10^{-4}
16	0.4590	0.6755	0.67551809	1.8089×10^{-5}	1.9600×10^{-4}	2.3000×10^{-4}	2.0600×10^{-4}
17	0.4784	0.6320	0.63102588	9.7411×10^{-4}	1.1090×10^{-3}	1.2710×10^{-3}	1.1170×10^{-3}
18	0.4960	0.5730	0.57300627	6.2700×10^{-6}	9.1027×10^{-4}	1.1120×10^{-3}	9.2000×10^{-4}
19	0.5119	0.4990	0.49898281	1.7190×10^{-5}	4.9902×10^{-4}	5.6300×10^{-4}	4.9000×10^{-4}
20	0.5265	0.4130	0.41270839	2.9160×10^{-4}	4.9030×10^{-4}	6.1200×10^{-4}	4.9200×10^{-4}
21	0.5398	0.3165	0.31629674	2.0325×10^{-4}	7.1532×10^{-4}	9.8500×10^{-4}	7.1800×10^{-4}
22	0.5521	0.2120	0.21218495	1.8495×10^{-4}	1.0468×10^{-4}	1.4200×10^{-4}	1.0200×10^{-4}
23	0.5633	0.1035	0.10350897	8.9700×10^{-6}	7.8397×10^{-4}	1.2540×10^{-3}	7.7900×10^{-4}
24	0.5736	−0.0100	−0.01025607	2.5607×10^{-4}	7.5437×10^{-4}	1.2680×10^{-3}	7.5100×10^{-4}
25	0.5833	−0.1230	−0.12309841	9.8410×10^{-5}	1.3775×10^{-3}	2.5370×10^{-3}	1.3810×10^{-3}
26	0.5900	−0.2100	−0.21005316	5.3159×10^{-5}	8.0320×10^{-4}	1.4690×10^{-3}	8.0700×10^{-4}

Table 6. Calculated current and compared IAE for double diode (RTC France Company).

Item	V_{Exp} (V)	I_{Exp} (A)	$I_{Calculated}$ (A)	Individual Absolute Error (IAE)		
				FA	RADE	NM-MPSO
1	−0.2057	0.7640	0.76404800	4.7990×10^{-5}	9.2680×10^{-5}	2.3000×10^{-5}
2	−0.1291	0.7620	0.76265838	6.5837×10^{-4}	6.5394×10^{-4}	5.9800×10^{-4}
3	−0.0588	0.7605	0.76138191	8.8191×10^{-4}	8.5755×10^{-4}	8.3200×10^{-4}
4	0.0057	0.7605	0.76020876	2.9123×10^{-4}	3.3747×10^{-4}	3.3000×10^{-4}
5	0.0646	0.7600	0.75912329	8.7671×10^{-4}	9.4000×10^{-4}	8.9500×10^{-4}
6	0.1185	0.7590	0.75806245	9.3754×10^{-4}	9.4935×10^{-4}	8.8000×10^{-4}
7	0.1678	0.7570	0.75700411	4.1100×10^{-6}	9.6350×10^{-5}	1.8700×10^{-4}
8	0.2132	0.7570	0.75750201	5.0201×10^{-4}	8.5535×10^{-4}	7.5700×10^{-4}
9	0.2545	0.7555	0.75557754	7.7540×10^{-5}	4.1885×10^{-4}	3.2300×10^{-4}
10	0.2924	0.7540	0.75409595	9.5950×10^{-5}	3.3126×10^{-4}	2.7700×10^{-4}
11	0.3269	0.7505	0.75031932	1.8060×10^{-4}	8.9511×10^{-4}	8.9600×10^{-4}
12	0.3585	0.7465	0.74651818	1.8185×10^{-5}	8.4939×10^{-4}	7.9800×10^{-4}
13	0.3873	0.7385	0.73873379	2.3370×10^{-4}	1.6021×10^{-3}	1.4950×10^{-3}
14	0.4137	0.7280	0.72816539	1.6540×10^{-4}	6.1216×10^{-4}	7.2900×10^{-4}
15	0.4373	0.7065	0.70628557	2.1442×10^{-4}	4.5162×10^{-4}	3.4400×10^{-4}
16	0.4590	0.6755	0.67594242	4.4242×10^{-4}	1.9888×10^{-4}	2.5900×10^{-4}
17	0.4784	0.6320	0.63286049	8.6045×10^{-4}	1.1123×10^{-3}	1.0990×10^{-3}
18	0.4960	0.5730	0.57381689	8.1689×10^{-4}	9.2523×10^{-4}	8.4500×10^{-4}
19	0.5119	0.4990	0.49879214	2.0785×10^{-4}	4.9417×10^{-4}	5.8600×10^{-4}
20	0.5265	0.4130	0.41276355	2.3644×10^{-4}	4.9125×10^{-4}	5.7100×10^{-4}
21	0.5398	0.3165	0.31674212	2.4212×10^{-4}	7.1918×10^{-4}	7.5300×10^{-4}
22	0.5521	0.2120	0.21202519	2.5196×10^{-5}	1.0831×10^{-4}	8.8000×10^{-5}
23	0.5633	0.1035	0.10350359	3.5935×10^{-6}	7.7968×10^{-4}	8.2700×10^{-4}
24	0.5736	−01000	−0.01049021	4.9021×10^{-4}	7.5539×10^{-4}	7.1100×10^{-4}
25	0.5833	−0.1230	−0.12300588	5.8808×10^{-6}	1.3767×10^{-3}	1.3880×10^{-3}
26	0.5900	−0.2100	−0.21005362	5.3621×10^{-5}	8.0501×10^{-4}	8.6500×10^{-4}

Furthermore, to understand the quality of the curve fit between Firefly algorithm values and experimental data, the results are compared to other algorithms. The compared statistical analysis for each model is presented in Tables 7 and 8. The compared statistical criteria indicates that the Firefly algorithm ranks the overall lowest values for relative error (RE), median absolute error (MAE), residual sum of squares (SSE), and root mean square error (RMSE). The statistical errors are used to show the performance with the definitions as follows:

$$RE = \frac{I_{measured} - I_{estimated}}{I_{measured}} \quad (17)$$

$$MAE = \sum_{i=1}^{m} \frac{|I_{\text{estimated}} - I_{\text{measured}}|}{m} \qquad (18)$$

$$SSE = \sum_{i=1}^{m} (I_{\text{measured}} - I_{\text{estimated}})^2 \qquad (19)$$

$$RMSE = \sqrt{\frac{1}{m} \sum_{i=1}^{m} (I_{\text{measured}} - I_{\text{estimated}})^2} \qquad (20)$$

Table 7. Statistical result for single diode model (RTC France Company).

Item	FA	BBO-M	RADE	LMSA	CARO	ABC	NM-MPSO
Total IAE	9.92230×10^{-3}	21.3000×10^{-3}	17.7036×10^{-3}	21.5104×10^{-3}	18.1550×10^{-3}	20.5000×10^{-3}	17.700×10^{-3}
RMSE	5.138165×10^{-4}	9.8634×10^{-4}	9.8602×10^{-4}	9.8640×10^{-4}	9.86650×10^{-4}	9.86200×10^{-4}	9.8602×10^{-4}
SSE	5.723673×10^{-6}	2.52997×10^{-5}	1.5625×10^{-5}	2.5297×10^{-5}	1.65385×10^{-5}	25.7000×10^{-6}	15.6295×10^{-6}
MAE	3.81630×10^{-4}	8.1923×10^{-4}	6.8090×10^{-4}	8.2732×10^{-4}	6.98260×10^{-4}	7.8846×10^{-4}	6.8077×10^{-4}

IAE: Individual Absolut Error, RMSE: Root Mean Square Error, SSE: Sum of Squares Error, MAE: Mean Absolute Error.

Table 8. Statistical result for double diode model (RTC France Company).

Item	FA	RADE	CARO	ABSO	ABC	NM-MPSO
Total IAE	8.570300×10^{-3}	17.7093×10^{-3}	69.330×10^{-3}	17.768×10^{-3}	20.3929×10^{-3}	17.356×10^{-3}
RMSE	4.548499×10^{-6}	9.82480×10^{-4}	9.8260×10^{-4}	9.8344×10^{-4}	9.8610×10^{-4}	9.8250×10^{-4}
SSE	5.379100×10^{-6}	15.6338×10^{-6}	16.9587×10^{-6}	15.3457×10^{-6}	25.600×10^{-6}	14.9455×10^{-6}
MAE	3.2963×10^{-4}	17.7093×10^{-3}	69.330×10^{-3}	17.768×10^{-3}	20.3929×10^{-3}	6.6754×10^{-4}

From Tables 7 and 8, we observe that the five and seven electrical parameters identified by the Firefly algorithm are close and more accurate than those found by all other compared algorithms. The performance of the proposed algorithm provides the lowest values for the statistical criteria, IAE, RMSE, SSE and MAE when compared to the other methods. Therefore, the Firefly algorithm is ranked first in achieving the lowest IAE, RMSE, SSE and MAE, while the Repaired Adaptive Differential Evolution algorithm and hybrid Nelder-Mead and Modified Particle Swarm Optimization (NM-MPSO) are ranked second and third, respectively. Therefore, the optimal parameters identified by the proposed Firefly Algorithm are very accurate because they are close to the real parameters of the system. The individual absolute error (IAE) and the relative error RE for each measurement using optimal values founded by the Firefly algorithm are illustrated in Figures 5 and 6, respectively. The Firefly algorithm performs better than the reported methods.

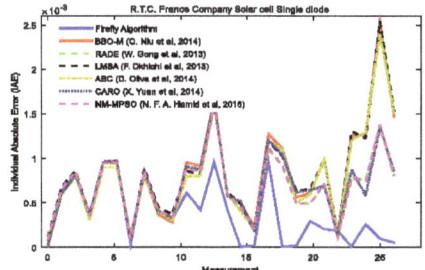
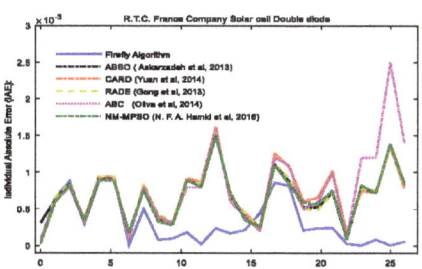

Figure 5. Individual absolute error (IAE) plots for single and double diode for Mono-crystalline silicon solar cell, RTC France Company.

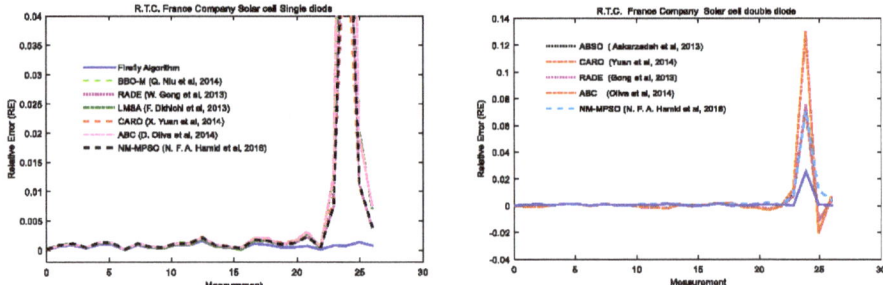

Figure 6. Relative Error (RE) plots for single and double diode for Mono-crystalline silicon solar cell, RTC France Company.

The current-voltage and power-voltage characteristics resulting from extracted parameters by the Firefly algorithm along with experimental data are compared to estimated data to investigate the quality of the identified parameters. This is illustrated in Figures 7 and 8. The two figures show the reconstructed single diode model is in good agreement with experimental data and are very close to each other.

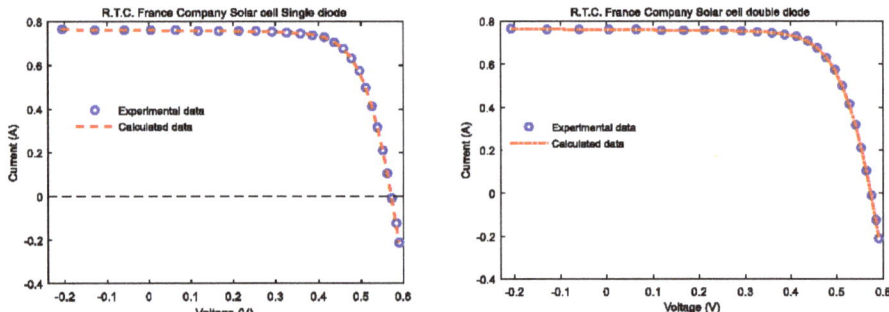

Figure 7. Experimental current-voltage data compared with estimated data of the ono-crystalline silicon solar cell single diode, RTC France Company.

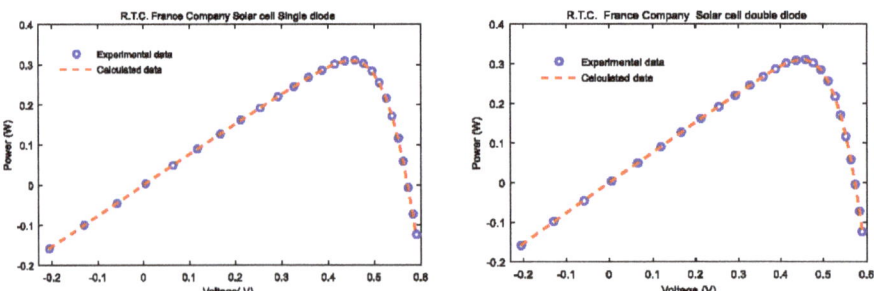

Figure 8. Experimental power-voltage data compared with estimated data of the mono-crystalline silicon solar cell single double diode, RTC France Company.

Figure 9 shows the compared extracted current-voltage characteristics of the mono-crystalline for single and double diode, RTC France Company. The calculated current by extracted parameters compared with the Firefly algorithm show good performance with the experimental data for single and double diode.

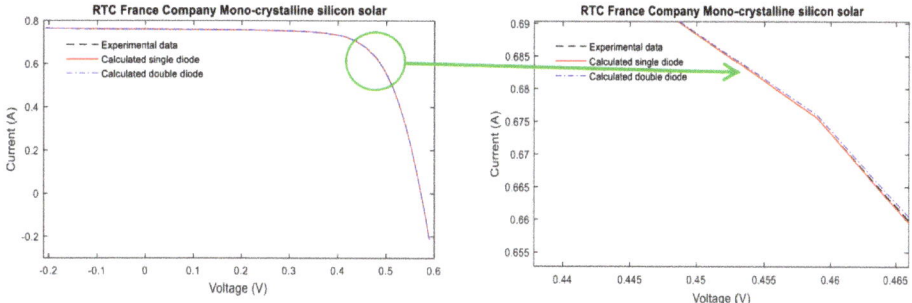

Figure 9. Compared experimental current-voltage and power-voltage of the mono-crystalline single diode silicon solar cell, RTC France Company.

5.2. Case 2: Organic Flexible Hydrogenated Amorphous Silicon a-Si:H Solar Cell

The Firefly algorithm, based on a parameter estimation method is used to extract the eight optimal parameters of flexible dual junction amorphous silicon solar cell under standard sunlight, based on data obtained in light intensity of 1000 W/m² and at a temperature of 300 K. The experimental data are used from [48]; only the open circuit voltage V_{oc} and short circuit current I_{sc} are obtained. Moreover, the optimal parameters are compared with several other techniques based on the same experimental data. The extracted optimal parameters by Firefly algorithm have been reported in Table 9, compared with the Quasi-Newton method and Self-Organizing Migrating Algorithm. Since it is difficult to extract the flexible amorphous silicon solar cell circuit model parameters and the research is still comparatively rare, the Quasi-Newton (Q-N) method and Self-Organizing Migrating Algorithm (SOMA) [48] have been chosen for comparison because in [29,48] they were demonstrated to provide good results for parameter extractions.

Table 9. Comparison among different parameter extraction of flexile silicon a-Si:H solar cell.

Algorithm	I_{ph} (µA)	d (m)	$\mu\tau_{eff}$ ($\frac{cm^2}{V}$)	V_{bi} (V)	R_s (Ω)	I_0 (A)	a	R_{sh} (Ω)
FA	0.3167	5.8065×10^{-8}	3.3306×10^{-5}	0.9895	0.4242	3.0691×10^{-14}	2	13.4978
Q-N	0.3043	5.8065×10^{-8}	4.8812×10^{-5}	0.9759	0.4242	3.0691×10^{-14}	1.9998	11.9138
SOMA	0.3181	4.9743×10^{-8}	3.3277×10^{-5}	0.9963	0.4706	3.0783×10^{-14}	1.9931	13.9288

To verify and validate the performance of the quality of the results, statistical analyses were carried out to measure the accuracy of the estimated parameters. The estimated current values are compared to experimental current by means of the following statistical errors: the individual absolute error (IAE), Standard deviation (SD), residual sum of squares (SSE), the root mean square error (RMSE) and the mean bias error (MBE) of the solar cell for each measurement, respectively. The statistical errors are used to compare term by term, the difference between estimated and experimental electric current. Generally, the lower these parameters, the more the efficiency of the model. Table 10 presents the current calculated for the Firefly algorithm and the individual absolute error, Table 11 summarizes the statistical errors for each measurement using the optimal values of x found by the Quasi-Newton method and Self-Organizing Migrating algorithm [48] compared with Firefly algorithm.

Table 10. Comparison between the calculate results of flexile silicon a-Si:H solar cell.

Experiment Current	FA		Q-N		SOMA	
	Current (A)	IAE	Current (A)	IAE	Current (A)	IAE
0	7.3656×10^{-4}	7.3656×10^{-4}	0.0041	0.0041	8.6804×10^{-4}	8.6804×10^{-4}
0.0158	0.0152	6.0×10^{-4}	0.0100	0.0058	0.0131	0.0027
0.0302	0.0361	0.0059	0.0305	0.0003	0.0334	0.0032
0.0619	0.0653	0.0034	0.0591	0.0028	0.0623	0.0004
0.0868	0.0744	0.0124	0.0680	0.0188	0.0715	0.0153
0.1142	0.1023	0.0119	0.0955	0.0187	0.1004	0.0138
0.1604	0.1623	0.0019	0.1549	0.0055	0.1679	0.0075
0.3044	0.3002	0.0042	0.2835	0.0209	0.3018	0.0026

Table 11. Performance indexes of flexile silicon a-Si:H solar cell.

Statistical Errors	FA	Q-N	SOMA
Standard deviation (SD)	4.925×10^{-3}	8.46×10^{-3}	7.86×10^{-3}
Root mean square error (RMSE)	6.1634×10^{-3}	12.3924×10^{-3}	7.9529×10^{-3}
Residual sum of squares (SSE)	3.6384×10^{-4}	1.2286×10^{-3}	5.0604×10^{-4}
Mean bias error (MBE)	6.62401×10^{-3}	1.2424×10^{-2}	7.4912×10^{-3}

Figure 10 presents the compared individual absolute error of each measurement used for current and power of optimal value x found by Firefly algorithm compared with the Quasi-Newton method and Self-Organizing Migrating Algorithm. From Figure 10, Tables 10 and 11 we know that the Firefly algorithm and Self-Organizing Migration Algorithm have the lowest SD, RMSE, SSE and MBE values among these three compared methods. Furthermore, the Firefly algorithm has better performance than the Quasi-Newton method and Self-Organizing Migration presented in [48].

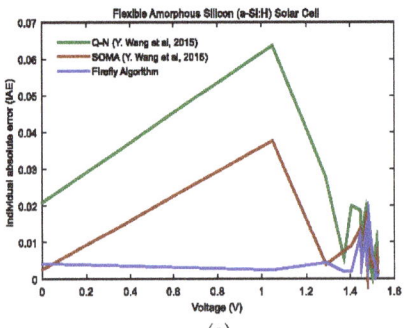
(a)

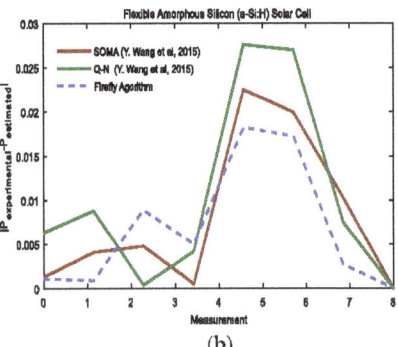
(b)

Figure 10. Individual absolute error compared to, (**a**) *I-V* and (**b**) *P-V* for each current measurement by different algorithms.

In order to illustrate the quality of the extracted optimal values x_3 found by the Firefly algorithm, the extracted values of I_{ph}, d_i, $\mu\tau$, V_{bi}, R_s, I_0, a and R_{sh} are put into Equation (3), then the current-voltage and power-voltage characteristics of this model is reconstructed with 16 pairs of current-voltage. The current-voltage and power-voltage characteristics resulting from the extracted parameters by Firefly algorithm along with experimental data have been illustrated in Figure 11. The Figures show the reconstructed model is in good agreement with the experimental data.

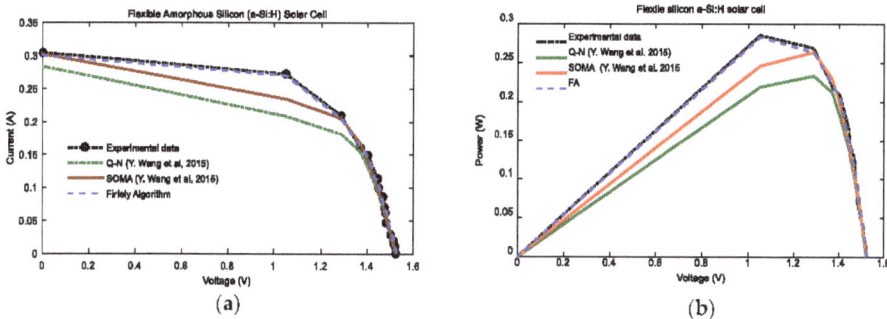

Figure 11. Comparison between, (**a**) *I-V* and (**b**) *P-V* characteristics resulting from the experimental data, Q-N, Soma and FA.

The comparative statistical error used in this paper compare the difference between estimated and experimental electric parameters, term by term. The mean bias error (MBE) provides information on the overestimation or underestimation of the solar cell performance. Therefore, the obtained results are more accurate than those found by Q-N and SOMA, can better reveal the actual behavior of solar cells and the model is efficient. Generally, the lower these parameters are, the more the model is efficient.

5.3. Case 3: Commercial Silicon Photovoltaic Module Photowatt-PWP 201

The prototype of the Photowatt-PWP 201 photovoltaic module has six solar panels, two are connected in series and three photovoltaic panels are connected in parallel. The measured voltage and current are taken under 25 °C and 1000 W/m². In this case, 26-pair current-voltage measured values are the same as [29], which are derived from 36 polycrystalline silicon cells which are connected in series. The extracted optimal parameters values for the photovoltaic module by Firefly algorithm have been reported in Table 12. Moreover, the optimal parameters are compared with several other techniques: Newton-Raphson [29] Pattern Search (PS) [55], Genetic Algorithm (GA) [56] and Simulated Annealing algorithm (SA) [58] based on the same experimental data. The purpose of comparison is to validate the accuracy of the Firefly algorithm in the parameter extraction process with a short time of convergence.

Table 12. Optimal parameter values identified by FA for Photowatt-PWP 201 polycrystalline photovoltaic module single diode compared with other methods.

Item	FA	Newton-Raphson	PS	GA	SA	NM-MPSO
I_{ph} (A)	1.0306	1.0318	1.0313	1.0441	1.0331	1.0305
I_0 (µA)	3.4802	3.2875	3.1756	3.4360	3.6642	3.6817
a	48.6551	48.4500	48.2889	48.5862	48.8211	48.8598
R_s (Ω)	1.2014	1.2057	1.2053	1.1968	1.1989	1.1944
R_{sh} (Ω)	971.1396	555.5556	714.2857	555.5556	833.3333	983.9970

The quality of the results in the extracted parameters are used to calculate the theoretical current values and compared to experimental measurements as show in Table 13.

The optimal value of the following statistical errors: individual absolute error (IAE), relative error (RE), root means square error (RMSE) and residual sum of squares (SSE), for each measurement using the Firefly algorithm and other parameter extraction techniques are given in Table 14.

Table 13. Measured and calculated current of photovoltaic module Photowatt-PWP 201 at 25 different working conditions compared with SA and PS.

Item	V_{Exp} (V)	I_{Exp} (A)	$I_{Calculated}$ (A)	Individual Absolute Error		
				FA	SA	PS
1	0.1248	1.0315	1.02919209	2.30790×10^{-3}	6.0000×10^{-5}	2.2000×10^{-3}
2	1.8093	1.0300	1.02743525	2.56480×10^{-3}	6.4000×10^{-4}	3.7800×10^{-3}
3	3.3511	1.0260	1.02577555	2.24450×10^{-4}	1.4100×10^{-3}	2.6500×10^{-3}
4	4.7622	1.0220	1.02412139	2.12140×10^{-3}	3.4900×10^{-3}	1.4100×10^{-3}
5	6.0538	1.0180	1.02228609	4.28610×10^{-3}	5.4100×10^{-3}	2.4000×10^{-4}
6	7.2364	1.0155	1.01990640	4.40640×10^{-3}	5.2900×10^{-3}	1.0100×10^{-3}
7	8.3189	1.0140	1.01632679	2.32680×10^{-3}	2.9600×10^{-3}	3.8800×10^{-3}
8	9.3097	1.0100	1.01045436	4.54360×10^{-4}	830.00×10^{-6}	6.4200×10^{-3}
9	10.2163	1.0035	1.00062757	2.87240×10^{-3}	2.8200×10^{-3}	10.320×10^{-3}
10	11.0449	0.9880	0.98458550	3.41450×10^{-3}	3.7000×10^{-3}	11.260×10^{-3}
11	11.8018	0.9630	0.95960866	3.39130×10^{-3}	4.0300×10^{-3}	11.450×10^{-3}
12	12.4929	0.9255	0.92293341	2.56660×10^{-3}	3.5000×10^{-3}	10.590×10^{-3}
13	13.1231	0.8725	0.87243997	6.00000×10^{-5}	1.0000×10^{-3}	7.5600×10^{-3}
14	13.6983	0.8075	0.80712359	3.76410×10^{-4}	1.5200×10^{-3}	7.4200×10^{-3}
15	14.2221	0.7265	0.72772952	1.22950×10^{-3}	4.4000×10^{-4}	4.7100×10^{-3}
16	14.6995	0.6345	0.63619518	1.69520×10^{-3}	1.2200×10^{-3}	3.0900×10^{-3}
17	15.1346	0.5345	0.53538376	8.83760×10^{-4}	3.6000×10^{-4}	3.0700×10^{-3}
18	15.5311	0.4275	0.42846560	9.65600×10^{-4}	8.0000×10^{-4}	1.7300×10^{-3}
19	15.8929	0.3185	0.31828380	2.16190×10^{-4}	7.4000×10^{-4}	2.3400×10^{-3}
20	16.2229	0.2085	0.20744219	1.05780×10^{-3}	1.8900×10^{-3}	2.5500×10^{-3}
21	16.5241	0.1010	0.09791334	3.08670×10^{-3}	5.3400×10^{-3}	5.0500×10^{-3}
22	16.7987	−0.008	−0.00863233	6.32300×10^{-4}	5.9000×10^{-4}	6.7000×10^{-4}
23	17.0499	−0.111	−0.11145028	4.50280×10^{-4}	6.0000×10^{-5}	2.2800×10^{-3}
24	17.2793	−0.209	−0.20961535	6.15350×10^{-4}	0000000000	3.1900×10^{-3}
25	17.4885	−0.303	−0.30253352	4.66470×10^{-4}	2.6200×10^{-3}	6.7500×10^{-3}

Table 14. Comparison of performance indexes for photovoltaic module Photowatt-PWP 201.

Item	FA	Newton-Raphson	PS	GA	SA
Total IAE	42.6725×10^{-3}	56.8800×10^{-3}	115.610×10^{-3}	153.479×10^{-3}	50.710×10^{-3}
RMSE	2.1540×10^{-3}	780.500×10^{-3}	11.8000×10^{-3}	6.9828×10^{-3}	2.700×10^{-3}
SSE	1.1600×10^{-4}	2.3249×10^{-4}	8.1725×10^{-4}	1.2190×10^{-3}	1.7703×10^{-4}
MAE	1.7069×10^{-3}	2.2752×10^{-3}	4.6244×10^{-3}	6.1392×10^{-3}	2.0284×10^{-3}

Table 14 proves that the Firefly algorithm has the lowest IAE, RMSE, SSE and MAE compared to other parameter extraction techniques such as, Newton-Raphson, Pattern Search (PS), Genetic Algorithm (GA) and Simulated Annealing algorithm (SA), since the Firefly algorithm found the minimum value of statistical analysis in parameter extraction for the photovoltaic module.

The comparison between Newton-Raphson, Pattern Search (PS), Genetic Algorithm (GA) and Simulated Annealing algorithm (SA) and the proposed algorithm, with the optimal value of IAE for each measurement, is illustrated in Figure 12. This Figure shows that the FA algorithm has better performance than the other parameter extraction algorithms. The total IAE values for each measurement is also calculated and listed in Table 14. The total IAE value shown in Table 14 highlights that the FA has the lowest total IAE compared to other algorithms for the photovoltaic module. Table 14 and Figure 12 indicate that FA outperforms the compared algorithms for this parameter extraction problem.

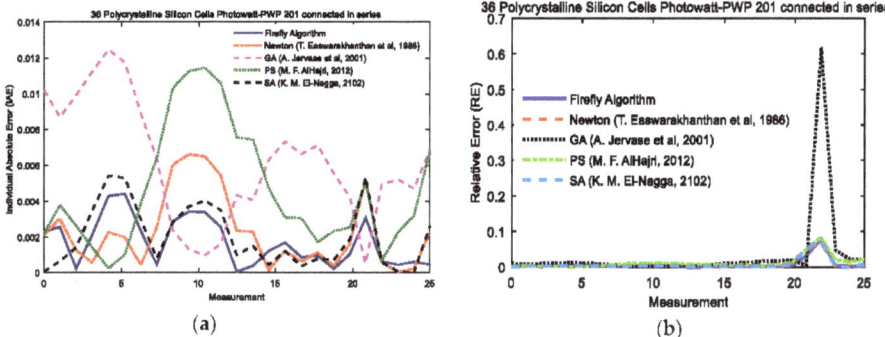

Figure 12. Comparison between, (**a**) IAE and (**b**) RE using the extracted parameters by FA and Newton-Raphson, PS, GA and SA for photovoltaic module Photowatt-PWP 201.

In order to validate the optimal values I_{ph}, I_0, a, R_s and R_p extracted by the Firefly algorithm, they are substituted into Equation (1) to reconstruct the current-voltage and power-voltage of the photovoltaic module. Figure 13 illustrates the current-voltage characteristics of the optimal values extracted by FA along with the experimental data. From the results, it can be observed that the values extracted by FA for the considered photovoltaic module fit the experimental data very well.

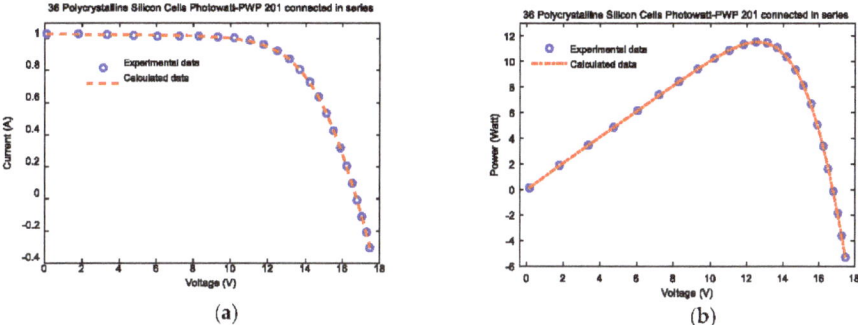

Figure 13. Comparison of (**a**) I-V and (**b**) P-V curve between experimentally recorded data for photovoltaic module Photowatt-PWP 201 and the estimated results by FA.

From these two cases for the solar cell, the single diode and double diode models, the Firefly algorithm showed the lowest statistical criteria: IAE, RMSE, SSE and MAE values among the compared techniques. We observed that the proposed algorithm is able to extract the intrinsic electrical parameters at the entire range of irradiance and temperature and performance, as compared to other recent techniques.

6. Conclusions

The paper presents the application of the Firefly algorithm in order to provide an accurate model of solar cells, single and double, and photovoltaic modules. The data required for testing the effectiveness of the Firefly algorithm optimization technique is based on the results in previous literature, experimental data and the nonlinear function of solar cell/photovoltaic characteristics. From the results and the statistical analyses, it can be observed that the proposed Firefly algorithm achieves the least root mean square error (RMSE), residual sum of squares (SSE) and mean absolute error (MAE) comparing the estimated and experimental data. Furthermore, the reproduction of current-voltage characteristics predicted using the parameters extracted by the Firefly algorithm are very close to those based on the experimental data. Moreover, the Firefly algorithm can extract the optimal parameters at all ranges of irradiance and temperature, especially at low irradiance.

Author Contributions: In this research activity, all of the authors were involved in the data analysis and preprocessing phase, the simulation, the results analysis and discussion, and the manuscript's preparation. All of the authors have approved the submitted manuscript. All the authors equally contributed to the writing of the paper.

Conflicts of Interest: The authors declare no conflict of interest.

References

1. Sen, S.; Ganguly, S. Opportunities, barriers and issues with renewable energy development a discussion. *Renew. Sustain. Energy Rev.* **2017**, *69*, 1170–1181. [CrossRef]
2. Tran, T.T.D.; Smith, A.D. Evaluation of renewable energy technologies and their potential for technical integration and cost-effective use within the U.S. energy sector. *Renew. Sustain. Energy Rev.* **2017**, *80*, 1372–1388. [CrossRef]
3. Karatepe, E.; Boztepe, M.; Çolak, M. Development of a suitable model for characterizing photovoltaic arrays with shaded solar cells. *Sol. Energy* **2007**, *81*, 977–992. [CrossRef]
4. Fathabadi, H. Novel neural-analytical method for determining silicon/plastic solar cells and modules characteristics. *Energy Convers. Manag.* **2013**, *76*, 253–259. [CrossRef]
5. Mares, O.; Paulescu, M.; Badescu, V. A simple but accurate procedure for solving the five-parameter model. *Energy Convers. Manag.* **2015**, *105*, 139–148. [CrossRef]
6. Pelap, F.B.; Dongo, P.D.; Kapim, A.D. Optimization of the characteristics of the PV cells using nonlinear electronic components. *Sustain. Energy Technol. Assess.* **2016**, *16*, 84–92.
7. Maghami, M.R.; Hizam, H.; Gomes, C.; Radzi, M.A.; Rezadad, M.I.; Hajighorbani, S. Power loss due to soiling on solar panel: A review. *Renew. Sustain. Energy Rev.* **2016**, *59*, 1307–1316. [CrossRef]
8. Fouad, M.M.; Shihata, L.A.; Morgan, E.I. An integrated review of factors influencing the performance of photovoltaic panels. *Renew. Sustain. Energy Rev.* **2017**, *80*, 1499–1511. [CrossRef]
9. Abidi, M.; Jabrallah, S.B.; Corriou, J.P. Optimization of the dynamic behavior of a solar distillation cell by Model Predictive Control. *Desalination* **2011**, *279*, 315–324.
10. Cotfas, D.T.; Cotfas, P.A.; Kaplanis, S. Methods and techniques to determine the dynamic parameters of solar cells: Review. *Renew. Sustain. Energy Rev.* **2016**, *61*, 213–221. [CrossRef]
11. Nemnes, G.A.; Besleaga, C.; Tomulescu, A.G.; Pintilie, I.; Pintilie, L.; Torfason, K.; Manolescu, A. Dynamic electrical behavior of halide perovskite based solar cells. *Sol. Energy Mater. Sol. Cells* **2017**, *159*, 197–203. [CrossRef]
12. Pavan, A.M.; Mellit, A.; Lughi, V. Explicit empirical model for general photovoltaic devices: Experimental validation at maximum power point. *Sol. Energy* **2014**, *101*, 105–116. [CrossRef]
13. Boutana, N.; Mellit, A.; Lughi, V.; Pavan, A.M. Assessment of implicit and explicit models for different photovoltaic modules technologies. *Energy* **2017**, *122*, 128–143. [CrossRef]
14. De Soto, W.; Klein, S.A.; Beckman, W.A. Improvement and validation of a model for photovoltaic array performance. *Sol. Energy* **2006**, *80*, 78–88. [CrossRef]
15. Brano, V.L.; Orioli, A.; Ciulla, G.; di Gangi, A. An improved five-parameter model for photovoltaic modules. *Sol. Energy Mater. Sol. Cells* **2010**, *94*, 1358–1370. [CrossRef]
16. Ding, K.; Zhang, J.; Bian, X.; Xu, J. A simplified model for photovoltaic modules based on improved translation equations. *Sol. Energy* **2014**, *101*, 40–52. [CrossRef]
17. Karmalkar, S.; Haneefa, S. A Physically Based Explicit J–V Model of a Solar Cell for Simple Design Calculations. *IEEE Electron Device Lett.* **2008**, *29*, 449–451. [CrossRef]
18. Das, A.K. An explicit J–V model of a solar cell for simple fill factor calculation. *Sol. Energy* **2011**, *85*, 1906–1909. [CrossRef]
19. Das, A.K. Analytical derivation of explicit J–V model of a solar cell from physics based implicit model. *Sol. Energy* **2012**, *86*, 26–30. [CrossRef]
20. Lun, S.; Wang, S.; Yang, G.; Guo, T. A new explicit double-diode modeling method based on Lambert W-function for photovoltaic arrays. *Sol. Energy* **2015**, *116*, 69–82. [CrossRef]

21. Boutana, N.; Mellit, A.; Haddad, S.; Rabhi, A.; Pavan, A.M. An explicit I-V model for photovoltaic module technologies. *Energy Convers. Manag.* **2017**, *138*, 400–412. [CrossRef]
22. Dehghanzadeh, A.; Farahani, G.; Maboodi, M. A novel approximate explicit double-diode model of solar cells for use in simulation studies. *Renew. Energy* **2017**, *103*, 468–477. [CrossRef]
23. Dolara, A.; Leva, S.; Manzolini, G. Comparison of different physical models for PV power output prediction. *Sol. Energy* **2015**, *119*, 83–99. [CrossRef]
24. Hasan, M.A.; Parida, S.K. An overview of solar photovoltaic panel modeling based on analytical and experimental viewpoint. *Renew. Sustain. Energy Rev.* **2016**, *60*, 75–83. [CrossRef]
25. Leva, S.; Dolara, A.; Grimaccia, F.; Mussetta, M.; Ogliari, E. Analysis and validation of 24 hours ahead neural network forecasting of photovoltaic output power. *Math. Comput. Simul.* **2017**, *131*, 88–100. [CrossRef]
26. Ogliari, E.; Dolara, A.; Manzolini, G.; Leva, S. Physical and hybrid methods comparison for the day ahead PV output power forecast. *Renew. Energy* **2017**, *113*, 11–21. [CrossRef]
27. Grimaccia, F.; Leva, S.; Mussetta, M.; Ogliari, E. ANN Sizing Procedure for the Day-Ahead Output Power Forecast of a PV Plant. *Appl. Sci.* **2017**, *7*, 622. [CrossRef]
28. Chen, Y.; Wang, X.; Li, D.; Hong, R.; Shen, H. Parameters extraction from commercial solar cells I–V characteristics and shunt analysis. *Appl. Energy* **2011**, *88*, 2239–2244. [CrossRef]
29. Easwarakhanthan, T.; Bottin, J.; Bouhouch, I.; Boutrit, C. Nonlinear Minimization Algorithm for Determining the Solar Cell Parameters with Microcomputers. *Int. J. Sol. Energy* **1986**, *4*, 1–12. [CrossRef]
30. Chan, D.S.H.; Phillips, J.R.; Phang, J.C.H. A comparative study of extraction methods for solar cell model parameters. *Solid-State Electron.* **1986**, *29*, 329–337. [CrossRef]
31. Ortiz-Conde, A.; Sánchez, F.J.G.; Muci, J. New method to extract the model parameters of solar cells from the explicit analytic solutions of their illuminated I–V characteristics. *Sol. Energy Mater. Sol. Cells* **2006**, *90*, 352–361. [CrossRef]
32. Nassar-eddine, I.; Obbadi, A.; Errami, Y.; el fajri, A.; Agunaou, M. Parameter estimation of photovoltaic modules using iterative method and the Lambert W function: A comparative study. *Energy Convers. Manag.* **2016**, *119*, 37–48. [CrossRef]
33. Gao, X.; Cui, Y.; Hu, J.; Xu, G.; Yu, Y. Lambert W-function based exact representation for double diode model of solar cells: Comparison on fitness and parameter extraction. *Energy Convers. Manag.* **2016**, *127*, 443–460. [CrossRef]
34. Peng, L.; Sun, Y.; Meng, Z. An improved model and parameters extraction for photovoltaic cells using only three state points at standard test condition. *J. Power Sources* **2014**, *248*, 621–631. [CrossRef]
35. Cubas, J.; Pindado, S.; de Manuel, C. Explicit Expressions for Solar Panel Equivalent Circuit Parameters Based on Analytical Formulation and the Lambert W-Function. *Energies* **2014**, *7*, 4098–4115. [CrossRef]
36. Lim, L.H.I.; Ye, Z.; Ye, J.; Yang, D.; Du, H. A Linear Identification of Diode Models from Single I-V Characteristics of PV Panels. *IEEE Trans. Ind. Electron.* **2015**, *62*, 4181–4193. [CrossRef]
37. Lim, L.H.I.; Ye, Z.; Ye, J.; Yang, D.; Du, H. A linear method to extract diode model parameters of solar panels from a single I–V curve. *Renew. Energy* **2015**, *76*, 135–142. [CrossRef]
38. Tsuno, Y.; Hishikawa, Y.; Kurokawa, K. Modeling of the I–V curves of the PV modules using linear interpolation/extrapolation. *Sol. Energy Mater. Sol. Cells* **2009**, *93*, 1070–1073. [CrossRef]
39. Lun, S.; Guo, T.; Du, C. A new explicit I–V model of a silicon solar cell based on Chebyshev Polynomials. *Sol. Energy* **2015**, *119*, 179–194. [CrossRef]
40. Lun, S.; Du, C.; Guo, T.; Wang, S.; Sang, J.; Li, J. A new explicit I–V model of a solar cell based on Taylor's series expansion. *Sol. Energy* **2013**, *94*, 221–232. [CrossRef]
41. Lun, S.; Du, C.; Yang, G.; Wang, S.; Guo, T.; Sang, J.; Li, J. An explicit approximate I–V characteristic model of a solar cell based on padé approximants. *Sol. Energy* **2013**, *92*, 147–159. [CrossRef]
42. Lun, S.; Du, C.; Sang, J.; Guo, T.; Wang, S.; Yang, G. An improved explicit I–V model of a solar cell based on symbolic function and manufacturer's datasheet. *Sol. Energy* **2014**, *110*, 603–614. [CrossRef]
43. Louzazni, M.; Aroudam, E.H. An analytical mathematical modeling to extract the parameters of solar cell from implicit equation to explicit form. *Appl. Sol. Energy* **2015**, *51*, 165–171. [CrossRef]
44. Zhang, Y.; Gao, S.; Gu, T. Prediction of I-V characteristics for a PV panel by combining single diode model and explicit analytical model. *Sol. Energy* **2017**, *144*, 349–355. [CrossRef]
45. Pindado, S.; Cubas, J. Simple mathematical approach to solar cell/panel behavior based on datasheet information. *Renew. Energy* **2017**, *103*, 729–738. [CrossRef]

46. Tong, N.T.; Pora, W. A parameter extraction technique exploiting intrinsic properties of solar cells. *Appl. Energy* **2016**, *176*, 104–115. [CrossRef]
47. Dkhichi, F.; Oukarfi, B.; Fakkar, A.; Belbounaguia, N. Parameter identification of solar cell model using Levenberg–Marquardt algorithm combined with simulated annealing. *Sol. Energy* **2014**, *110*, 781–788. [CrossRef]
48. Askarzadeh, A.; Rezazadeh, A. Artificial bee swarm optimization algorithm for parameters identification of solar cell models. *Appl. Energy* **2013**, *102*, 943–949. [CrossRef]
49. Oliva, D.; Cuevas, E.; Pajares, G. Parameter identification of solar cells using artificial bee colony optimization. *Energy* **2014**, *72*, 93–102. [CrossRef]
50. Hamid, N.F.A.; Rahim, N.A.; Selvaraj, J. Solar cell parameters identification using hybrid Nelder-Mead and modified particle swarm optimization. *J. Renew. Sustain. Energy* **2016**, *8*, 015502. [CrossRef]
51. Louzazni, M.; Crăciunescu, A.; Dumitrache, A. Identification of Solar Cell Parameters with Firefly Algorithm. In Proceedings of the 2015 Second International Conference on Mathematics and Computers in Sciences and in Industry (MCSI), Sliema, Malta, 17–19 August 2015; pp. 7–12.
52. Louzazni, M.; Khouya, A.; Amechnoue, K.; Crăciunescu, A.; Mussetta, M. Comparative prediction of single and double diode parameters for solar cell models with firefly algorithm. In Proceedings of the 2017 10th International Symposium on Advanced Topics in Electrical Engineering (ATEE), Bucharest, Romania, 23–25 March 2017; pp. 860–865.
53. Louzazni, M.; Khouya, A.; Amechnoue, K. A firefly algorithm approach for determining the parameters characteristics of solar cell. *Leonardo Electron. J. Pract. Technol.* **2017**, *31*, 235–250.
54. Wang, Y.; Xi, J.; Han, N.; Xie, J. Modeling method research of flexible amorphous silicon solar cell. *Appl. Sol. Energy* **2015**, *51*, 41–46. [CrossRef]
55. AlHajri, M.F.; El-Naggar, K.M.; AlRashidi, M.R.; Al-Othman, A.K. Optimal extraction of solar cell parameters using pattern search. *Renew. Energy* **2012**, *44*, 238–245. [CrossRef]
56. Jervase, J.A.; Bourdoucen, H.; Al-Lawati, A. Solar cell parameter extraction using genetic algorithms. *Meas. Sci. Technol.* **2001**, *12*, 1922. [CrossRef]
57. Zagrouba, M.; Sellami, A.; Bouaïcha, M.; Ksouri, M. Identification of PV solar cells and modules parameters using the genetic algorithms: Application to maximum power extraction. *Sol. Energy* **2010**, *84*, 860–866. [CrossRef]
58. El-Naggar, K.M.; AlRashidi, M.R.; AlHajri, M.F.; Al-Othman, A.K. Simulated Annealing algorithm for photovoltaic parameters identification. *Sol. Energy* **2012**, *86*, 266–274. [CrossRef]
59. Gong, W.; Cai, Z. Parameter extraction of solar cell models using repaired adaptive differential evolution. *Sol. Energy* **2013**, *94*, 209–220. [CrossRef]
60. Ye, M.; Wang, X.; Xu, Y. Parameter extraction of solar cells using particle swarm optimization. *J. Appl. Phys.* **2009**, *105*, 094502. [CrossRef]
61. Askarzadeh, A.; Coelho, L.d.S. Determination of photovoltaic modules parameters at different operating conditions using a novel bird mating optimizer approach. *Energy Convers. Manag.* **2015**, *89*, 608–614. [CrossRef]
62. Wang, L.; Huang, C. A novel Elite Opposition-based Jaya algorithm for parameter estimation of photovoltaic cell models. *Opt. Int. J. Light Electron Opt.* **2018**, *155*, 351–356. [CrossRef]
63. Yang, X.-S. Multiobjective firefly algorithm for continuous optimization. *Eng. Comput.* **2013**, *29*, 175–184. [CrossRef]
64. Yang, X.-S. *Nature-Inspired Metaheuristic Algorithms*; Luniver Press: Frome, UK, 2010.
65. Yang, X.-S.; He, X.-S. Why the Firefly Algorithm Works? In *Nature-Inspired Algorithms and Applied Optimization*; Springer: Cham, Switzerland, 2018; pp. 245–259.
66. Yang, X.-S. *Engineering Optimization: An Introduction with Metaheuristic Applications*; Wiley: Hoboken, NJ, USA, 2010.
67. Fister, I.; Fister, I.; Yang, X.-S.; Brest, J. A comprehensive review of firefly algorithms. *Swarm Evol. Comput.* **2013**, *13*, 34–46. [CrossRef]
68. Niu, Q.; Zhang, L.; Li, K. A biogeography-based optimization algorithm with mutation strategies for model parameter estimation of solar and fuel cells. *Energy Convers. Manag.* **2014**, *86*, 1173–1185. [CrossRef]
69. Yuan, X.; He, Y.; Liu, L. Parameter extraction of solar cell models using chaotic asexual reproduction optimization. *Neural Comput. Appl.* **2015**, *26*, 1227–1239. [CrossRef]

70. Soon, J.J.; Low, K.S. Optimizing Photovoltaic Model for Different Cell Technologies Using a Generalized Multidimension Diode Model. *IEEE Trans. Ind. Electron.* **2015**, *62*, 6371–6380. [CrossRef]
71. Merten, J.; Asensi, J.M.; Voz, C.; Shah, A.V.; Platz, R.; Andreu, J. Improved equivalent circuit and analytical model for amorphous silicon solar cells and modules. *IEEE Trans. Electron Devices* **1998**, *45*, 423–429. [CrossRef]
72. Hubin, J.; Shah, A.V. Effect of the recombination function on the collection in a p-i-n solar cell. *Philos. Mag. Part B* **1995**, *72*, 589–599. [CrossRef]
73. Voswinckel, S.; Wesselak, V.; Lustermann, B. Behaviour of amorphous silicon solar modules: A parameter study. *Sol. Energy* **2013**, *92*, 206–213. [CrossRef]
74. Merten, J.; Andreu, J. Clear separation of seasonal effects on the performance of amorphous silicon solar modules by outdoor I/V-measurements. *Sol. Energy Mater. Sol. Cells* **1998**, *52*, 11–25. [CrossRef]
75. Shah, A.V.; Sculati-Meillaud, F.; Berényi, Z.J.; Ghahfarokhi, O.M.; Kumar, R. Diagnostics of thin-film silicon solar cells and solar panels/modules with variable intensity measurements (VIM). *Sol. Energy Mater. Sol. Cells* **2011**, *95*, 398–403. [CrossRef]
76. Nonomura, S.; Okamoto, H.; Hamakawa, Y. Determination of the Built-in Potential in a-Si Solar Cells by Means of Electroabsorption Method. *Jpn. J. Appl. Phys.* **1982**, *21*, L464–L466. [CrossRef]
77. Dongaonkar, S.; Karthik, Y.; Wang, D.; Frei, M.; Mahapatra, S.; Alam, M.A. On the Nature of Shunt Leakage in Amorphous Silicon p-i-n Solar Cells. *IEEE Electron Device Lett.* **2010**, *31*, 1266–1268. [CrossRef]
78. Dongaonkar, S.; Servaites, J.D.; Ford, G.M.; Loser, S.; Moore, J.; Gelfand, R.M.; Mohseni, H.; Hillhouse, H.W.; Agrawal, R.; Ratner, M.A.; et al. Universality of non-Ohmic shunt leakage in thin-film solar cells. *J. Appl. Phys.* **2010**, *108*, 124509. [CrossRef]
79. El-Fergany, A. Efficient Tool to Characterize Photovoltaic Generating Systems Using Mine Blast Algorithm. *Electr. Power Compon. Syst.* **2015**, *43*, 890–901. [CrossRef]

© 2018 by the authors. Licensee MDPI, Basel, Switzerland. This article is an open access article distributed under the terms and conditions of the Creative Commons Attribution (CC BY) license (http://creativecommons.org/licenses/by/4.0/).

Article

Comparison of Training Approaches for Photovoltaic Forecasts by Means of Machine Learning

Alberto Dolara [†], Francesco Grimaccia [†], Sonia Leva [†], Marco Mussetta [†] and Emanuele Ogliari [†,*]

Dipartimento di Energia, Politecnico di Milano, via La Masa 34, 20156 Milano, Italy; alberto.dolara@polimi.it (A.D.); francesco.grimaccia@polimi.it (F.G.); sonia.leva@polimi.it (S.L.); marco.mussetta@polimi.it (M.M.)
* Correspondence: emanuelegiovanni.ogliari@polimi.it; Tel.: +39-2399-8524
† These authors contributed equally to this work.

Received: 31 December 2017; Accepted: 28 January 2018; Published: 2 February 2018

Abstract: The relevance of forecasting in renewable energy sources (RES) applications is increasing, due to their intrinsic variability. In recent years, several machine learning and hybrid techniques have been employed to perform day-ahead photovoltaic (PV) output power forecasts. In this paper, the authors present a comparison of the artificial neural network's main characteristics used in a hybrid method, focusing in particular on the training approach. In particular, the influence of different data-set composition affecting the forecast outcome have been inspected by increasing the training dataset size and by varying the training and validation shares, in order to assess the most effective training method of this machine learning approach, based on commonly used and a newly-defined performance indexes for the prediction error. The results will be validated over a one-year time range of experimentally measured data. Novel error metrics are proposed and compared with traditional ones, showing the best approach for the different cases of either a newly deployed PV plant or an already-existing PV facility.

Keywords: photovoltaics; power forecasting; artificial neural networks

1. Introduction

In recent years, several forecasting methods have been developed for the output power of renewable energy sources (RES) [1], addressing in particular the intrinsic variability of parameters related to changing weather conditions, which directly affect the photovoltaic (PV) systems' power output [2]. This increasing attention is mainly due to the increasing shares of RES quota in power systems, which involve novel technical challenges for the efficiency of the electrical grid [3]. In particular, predictive tools based on historical data can generally provide advantages in PV plant operation [4,5], reduce excess production, and take advantage of incentives for RES production [6].

Among the commonly-used forecasting models, most aim to predict the expected power production based on numerical weather prediction (NWP) systems forecasts [7]. This is a complex problem with high degrees of non-linearity; for this reason, it is commonly approached by means of advanced models and techniques—i.e., evolutionary computation [8], machine learning (ML) [9], and artificial neural networks (ANNs) [10]. These are pseudo-stochastic iterative approaches defined in the class of computational intelligence techniques, and are usually employed to address pattern recognition, function approximation, control, and forecasting problems [11]. Moreover, they are generally able to handle incomplete or missing data and solve problems with a high degree of complexity.

Recently, several ANN layouts have been developed to solve different tasks [12], such as: times series prediction, complex dynamical system emulation [13], speech generation, handwritten digit recognition, and image compression, due to their ability to learn from extended time series of historical measurements with acceptable error levels compared to other statistical and physical forecasting models [14]. Currently, ANN employment in forecasting is quite straightforward due to the widespread development of specific software applications [15–17].

In particular, the first attempts at solar power forecasting by means of ANN started more than a decade ago [18]. Generally, in the case of PV power output, common training data are the historical measurements of power production from a PV facility and meteorological parameters unique to the facility location, including temperature, global horizontal irradiance (i.e., the intensity of all the solar radiation components on a horizontal surface) [19], and cloud cover above the facility. Additional forecasted variables from the numerical weather predictions can also be considered, such as wind speed, humidity, pressure, etc. [20].

Novel forecasting models were recently implemented by adding an estimate of the clear sky radiation to the series of historical local weather data, as reported in [21].

Additionally, the effectiveness of ensemble methods was demonstrated in [22], thus giving additional advantages in terms of results reliability and the implementation of efficient parallel computing techniques.

In their previous work [23], the authors conducted a detailed analysis to find a procedure for the best ANN layout and settings in terms of the number of layers, neurons, and trials for the PV day-ahead forecast. Furthermore, evidence showed that the forecasting performance of ML techniques is affected by the composition of the training data-set, as well as by input selection [24,25].

In this paper, a specific study is conducted on training data-sets in order to provide a more detailed analysis of the effect of different approaches in the training data-set composition on the day-ahead forecast of the PV power production. In particular, the authors present some procedures to set-up the training and validation data-sets for the ANN used in physical hybrid method to perform the day-ahead PV power forecast in view of the electricity market. Moreover, a novel error metric is proposed and compared with traditional ones, in order to validate the best training approach in different cases: indeed, the procedures outlined herein can be adopted to set-up data-sets based on either historical data retrieved from an existing PV plant or on incremental data measurements in a newly deployed PV plant. The test data set will be made up of the 24-hourly PV power values forecasted one day-ahead.

The paper is structured as follows: Section 2 provides an overview of the considered approaches for the composition of the training database, considering both cases of historical data retrieved from an existing PV plant and incremental data measurements in a newly deployed PV plant; Section 3 presents the methodology implemented to compare the different training approaches presented here, proposing some new metrics aimed at evaluating the suitability of the proposed configurations in terms of error performance and statistical behavior; Section 4 presents the considered case study, which is used to test the proposed training approaches: specific simulations and numerical results are provided in Section 5, and final remarks are reported in Section 6.

2. Training Database Composition Approaches

In order to perform the day-ahead forecast, the ANN needs to be trained. Hence, the amount of historical data employed in the supervised learning determines the ANN forecast capability. This amount of data is formed of samples exploited in the process of identifying the links among neurons in the network which minimize the error in the forecast. In order to do this task, the whole amount of available samples is divided in two groups:

- the "training set" (or equally "training database"), which is used to adjust the weights among neurons by performing the forecast on the same samples,
- the "validation set", which is used as a stopping criteria to avoid over-fitting and under-fitting. It proves the goodness of the trained network on additional samples which have not been previously included in the training set. The purpose of this step is to test the generalization capability of the neural network on a new data-set .

Learning occurs by updating elements within the network; thus, its response iteratively improves to match the desired output. An ANN is trained when it has learned its task and converges to a solution. To achieve this, some learning algorithms are commonly used:

- error back-propagation (EBP)
- gradient descent
- conjugate gradient
- evolutionary algorithms (genetic algorithms, particle swarm optimization, etc.)

Sometimes, according to the problem, the fastest algorithm gives solutions rapidly converging on local minima; however, this does not guarantee the maximum accuracy. In addition, it should be considered that a large training set size provides a better sample of the trends improving generalization, but it generally slows down the learning process. If an ANN is not properly trained or sized, there are usually undesired results, such as "overfitting" and "underfitting" [26]. Using ANN ensembles by averaging their outputs has been demonstrated to be beneficial, as it helps to avoid chance correlations and the overtraining problem [27,28].

However, to choose both the most suitable learning algorithm and the proper size of the training set which minimizes the error is a challenge which should be faced in each case study [29–31].

In this paper, we inspect how the behavior on the day-ahead forecast is influenced by the possible characteristics summarized in Figure 1. The first characteristic of the data-set is either "incremental" when the elements belonging to the training data-set are progressively available over time and the training set size gradually increases or "complete" if an already existing database of samples is available. The second characteristic refers to the way the data-set is used for training the ANN. As the forecast-making is mainly a stochastic process, the choice could be to use entirely the same training data set for each forecast of the ensemble (we refer to the single forecast with the term "trial", and in this case, all the trials will be the same in the ensemble) or to shuffle its elements, grouping them in smaller subsets adopted each time to separately train a different ANN (in this other case, each trial is independent, as all the training data-sets are different). Finally, the mean of the resulting output is usually calculated in the so-called "ensemble" forecast. The third characteristic is related to the order of the hourly samples that constitute the training data-set. They can appear either consecutively displaced as the chronological time series they belong to or they can be randomly grouped and mixed up.

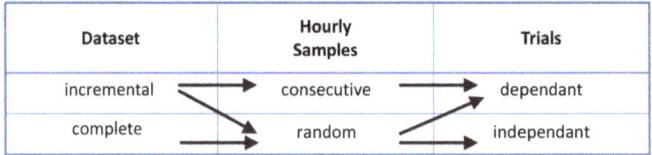

Figure 1. Main features of the ANN training data-sets.

The combination of these characteristics results in different ANN training methods, which could affect the forecast.

All of the assumptions exposed here are valid, in general terms, for all ANN-based methods. In this specific paper, authors employ the Physical Hybrid Artificial Neural Network (PHANN) method for the day-ahead forecast, as described in detail in [14,21]. This procedure mixes the physical Clear Sky Radiation Model (CSRM) and the stochastic ANN method as reported in Figure 2.

Appl. Sci. **2018**, *8*, 228

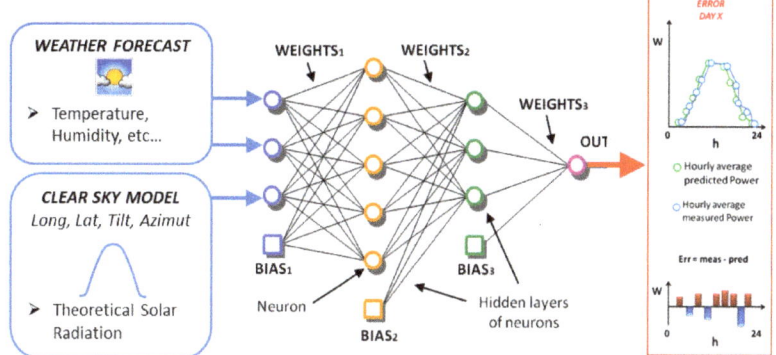

Figure 2. Physical Hybrid Artificial Neural Network (PHANN) method schematic diagram.

2.1. Incremental Training Data-Set

An incremental data-set occurs when the available samples are limited. Usually this is the case of real-time or time-dependant processes, and data can be acquired only progressively. Consider for example our case study when the monitoring system starts recording data from the first day of operation of the PV plant: initially a small amount of data is recorded, and if we acquire hourly samples, 24 samples are added to the historical data-set every day.

In this database composition (e.g., see Figure 3), the days which can be employed for ANN training are those available starting from the PV plant commissioning (day 1) until the k_d day before the forecast (day X_d). As a consequence, the size of the training database will increase over time. In order to supply the data-set to the network for the training step, samples can be arranged in different methods. Those adopted in this paper are listed in Figure 4, and determine different results in the forecast. A short description is given in the following:

- Method A employs the same chronologically consecutive samples by grouping the 90% of the samples which are closest to the forecast day for the training set and the remaining 10% of the samples for the validation set.
- Method A* employs the same chronologically consecutive samples by grouping the 90% of the samples for the training set and the 10% of the samples which are closest to the forecast day for the validation set.
- Method B employs the samples by randomly grouping them separately, 90% for the training set and 10% for the validation set.

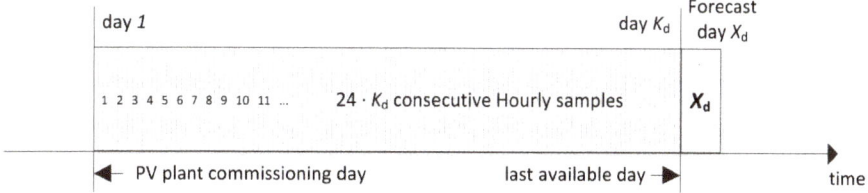

Figure 3. Hourly samples are progressively available in an incremental training database. PV: photovoltaic.

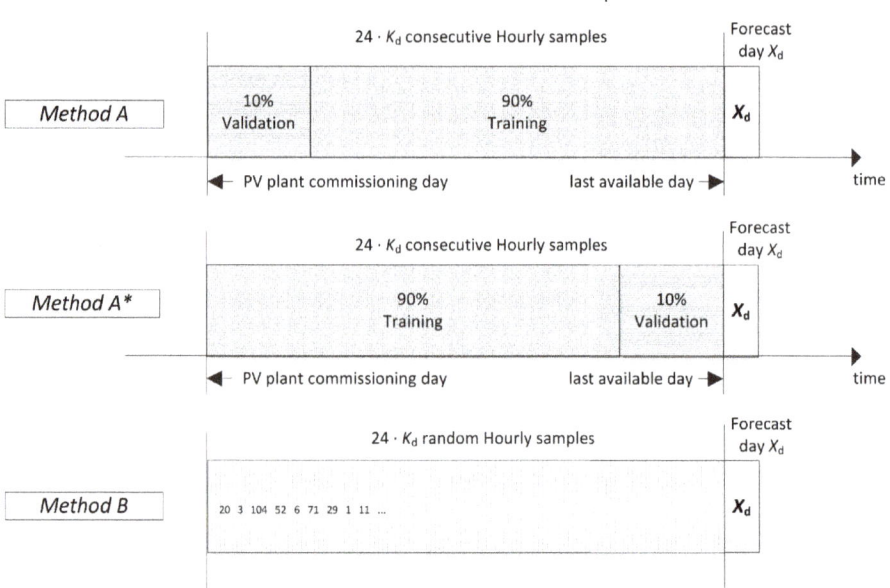

Figure 4. Training database composition for methods A, A*, and B.

In the first two methods (A and A*), the effect of the proximity of the training set to the forecast day is examined (implying seasonal variations on the parameter), inspecting how the forecast is affected by the proximity of the samples employed in the training rather than in the validation step. For example, it is clear that forecasting spring days cannot be accurate if the training samples belong to the past autumn or winter, and the same consideration applies for the validation. Reasonably, we are expecting that the further the samples of the validation are, the less accurate the forecast. Obviously, this problem is not addressed in Method B, as samples are randomly chosen.

2.2. Complete Training Data-Set

In the complete data-set, an extended amount of samples is available, but it might belong to a period of time which is time-wise distant from the days of the forecast, as it is shown in Figure 5. In this case, samples which have to be employed for the ANN training can either be mixed (as shown in Figure 6) each time that a trial is performed (this happens when trials are independent with Method C1), or each trial depends on the same training data-set with Method C2.

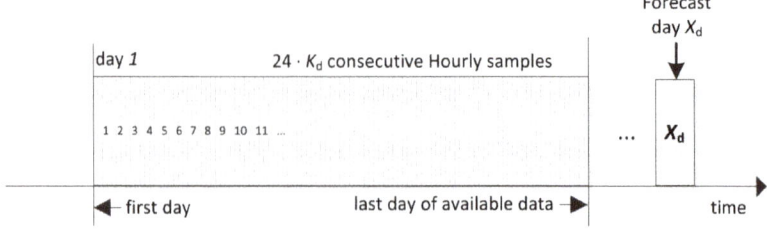

Figure 5. Hourly samples belonging to an extended period of time are available in a complete training database.

The complete list of the training methods which have been adopted in this paper is in Table 1. The different shares of the training and the validation set, 90% and 10%, respectively, have been set up in previous works.

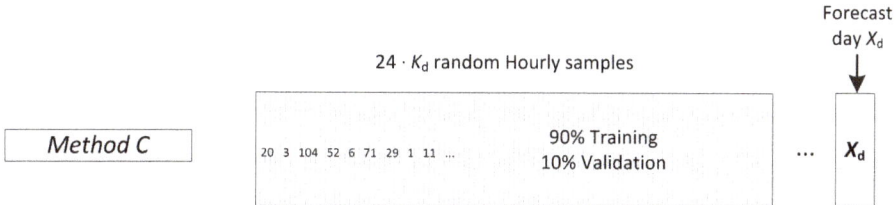

Figure 6. Hourly samples belonging to an extended period of time in a complete training database are randomly mixed.

Table 1. Different methods for the composition of the ANN training data-sets which have been analysed. † (90%*ts* 10%*vs*) *ts* = training set; *vs* = validation set.

Method	Data-Set	Trials	Samples
A	Incremental	Dependent	Consecutive (10%*vs* 90%*ts*)
A*	Incremental	Dependent	Consecutive (90%*ts* 10%*vs*)
B1	Incremental	Independent	Random †
B2	Incremental	Dependent	Random †
C1	Complete	Independent	Random †
C2	Complete	Dependent	Random †

3. Evaluation Indexes

The effect of the different methods of training is investigated by means of some evaluation indexes. These indexes aim at assessing the accuracy of the forecasts and the related error, and it is therefore necessary to define the indexes. There is a wide variety of existing definitions of the forecasting performance, and technical papers present many of these indexes; hence, we will report some of the most commonly used definitions in the literature ([32–34]).

The hourly error e_h is the starting definition given as the difference between the hourly mean values of the power measured in the h-th hour $P_{m,h}$ and the forecast $P_{p,h}$ provided by the adopted model [32,35]:

$$e_h = P_{m,h} - P_{p,h} \quad (W). \tag{1}$$

From the hourly error expression and its absolute value $|e_h|$, other definitions can be inferred; i.e., the well-known mean absolute percentage error ($MAPE$):

$$MAPE = \frac{1}{N} \sum_{h=1}^{N} \left| \frac{e_h}{P_{m,h}} \right| \cdot 100, \tag{2}$$

where N represents the number of samples (hours) considered: usually it is calculated for a single day, month, or year.

Since the hourly measured power $P_{m,h}$ significantly changes during the same day (i.e., sunrise, noon, and sunset), for the sake of a fair comparison, in this paper the authors preferred to consider the *normalized mean absolute error* $NMAE_\%$:

$$NMAE_\% = \frac{1}{N} \sum_{h=1}^{N} \left|\frac{e_h}{C}\right| \cdot 100, \tag{3}$$

where the percentage of the absolute error is referred to the rated power C of the plant, in place of the hourly measured power $P_{m,h}$.

In this paper we also adopted the mean value of all the $NMAE_{\%,d}$, which refers to the d-th day, calculated over the whole period. Therefore, we introduce $\overline{NMAE_\%}$, which is the mean of all the daily $NMAE_{\%,d}$ obtained with a given data-set:

$$\overline{NMAE_\%} = \frac{1}{D} \sum_{d=1}^{D} NMAE_{\%,d}. \tag{4}$$

The *weighted mean absolute error* $WMAE_\%$ is based on total energy production:

$$WMAE_\% = \frac{\sum_{h=1}^{N} |e_h|}{\sum_{h=1}^{N} P_{m,h}} \cdot 100. \tag{5}$$

The *normalized root mean square error* $nRMSE$ is based on the maximum hourly power output $P_{m,h}$:

$$nRMSE_\% = \frac{\sqrt{\frac{\sum_{h=1}^{N} |e_h|^2}{N}}}{max(P_{m,h})} \cdot 100. \tag{6}$$

This error definition is the well-known root mean square error ($RMSE$) which has been normalized over the maximum hourly power output $P_{m,h}$ measured in the considered time range, for the sake of a fair comparison.

$NMAE_\%$ is largely used to evaluate the accuracy of predictions and trend estimations. In fact, often relative errors are large because they are divided by small power values (for instance the low values associated to sunset and sunrise): in such cases, $WMAE_\%$ could result very large and biased, while $NMAE_\%$, by weighting these values with the capacity of the plant C, is more useful.

The $nRMSE_\%$ measures the mean magnitude of the absolute hourly errors $e_{h,abs}$. In fact, it gives a relatively higher weight to larger errors, thus allowing particularly undesirable results to be emphasized. In fact, if we consider the daily trends of the aforementioned indexes (which are shown in Figure 7), it can be seen how they are correlated, while in the same Figure 8, the scatterplot of their normalized values with the relative maxima clearly shows these correlations between the three error indexes. Furthermore, the Pearson–Bravais correlation index ρ_{xy} [36] has been calculated to underline the direct relationship among the error indexes:

$$\rho_{xy} = \frac{\sum_{h=1}^{N} (x_i - \mu_x)(y_i - \mu_y)}{\sqrt{\sum_{h=1}^{N} (x_i - \mu_x)^2} \sqrt{\sum_{h=1}^{N} (y_i - \mu_y)^2}}. \tag{7}$$

However, as it is shown in Figure 7, the daily evaluation indexes expressed in Equations (3), (5), and (6) could vary a great deal, being unable to give complete information "at a glimpse" of the accuracy of the prediction. For example, consider Figures 9 and 10, where the forecasts and the relevant evaluation indexes for 1 April and 4 November 2014, respectively, are depicted. In both cases, daily $NMAE_\%$ values are quite low (around 2–3%) and a forecast assessment solely based on this basis could be misleading.

Actually, the 1 April was quite a sunny day and the bell-shaped hourly power curve which has been forecast—the red starred line—was accurately following the measured one—the blue circled line. The cloudy winter day 4 November 2014 was a different story; in fact, the forecast red curve is biased on the noon hours, while the actual blue curve in the morning. However, in the second day, the daily $NMAE_\%$ value is lower. This is owing to the normalisation of the mean absolute error with the net capacity of the plant. Regarding the other evaluation indexes, even if they are correlated, they can exceed the 100% cap, as happens for example to $WMAE_\%$ in Figure 7 on day 72.

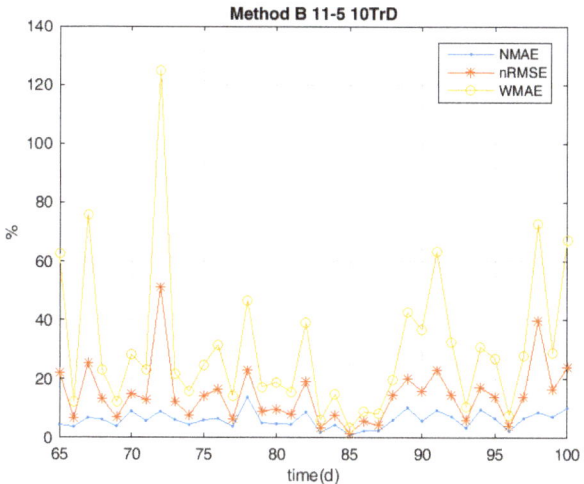

Figure 7. Example of the daily errors trend. $NMAE$: normalized mean absolute error; $nRMSE$: normalized root mean square error; $WMAE$: weighted mean absolute error.

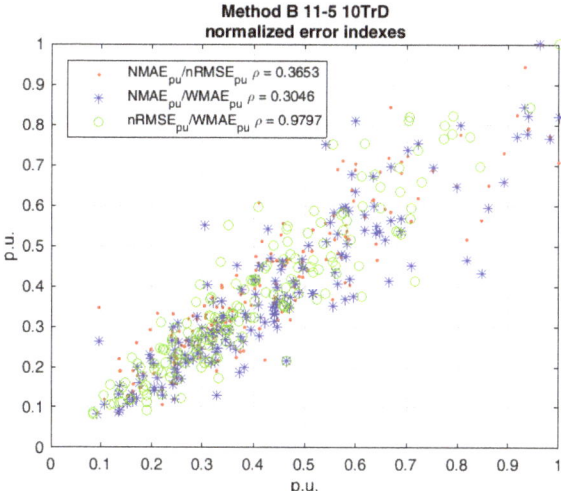

Figure 8. Normalized daily errors correlated in a scatterplot.

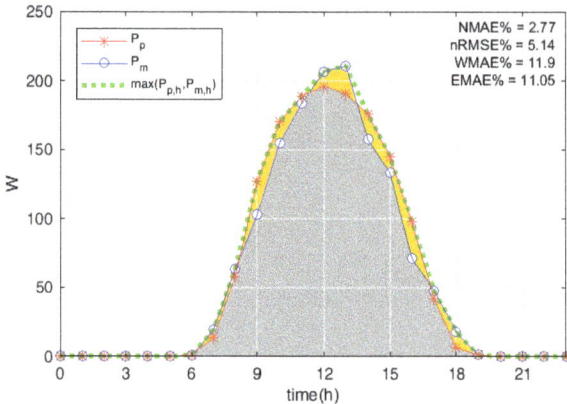

Figure 9. Example of a sunny day forecast—1 April 2014—with the relevant evaluation indexes. $EMAE\%$: envelope-weighted mean absolute error.

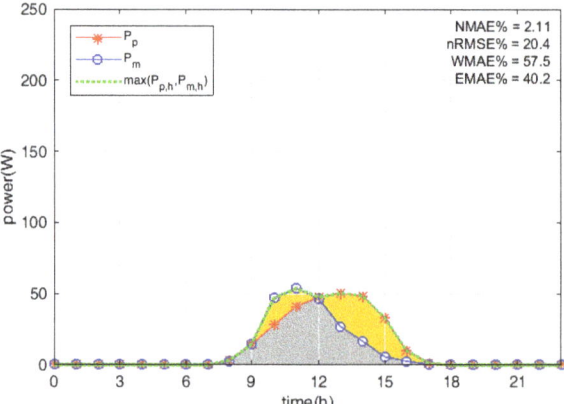

Figure 10. Example of a cloudy day forecast—4 November 2014—with the relevant evaluation indexes.

Starting from these assumptions, and in view of a more useful summary evaluation, an additional performance index is proposed, aiming to provide a value between 0% and 100% of the forecast accuracy. Therefore the *envelope-weighted mean absolute error*, $EMAE_\%$ is defined as:

$$EMAE_\% = \frac{\sum_{h=1}^{N} |e_h|}{\sum_{h=1}^{N} max(P_{m,h}, P_{p,h})} \cdot 100, \qquad (8)$$

where the numerator is the sum of the absolute hourly errors, as in $WMAE_\%$, while the denominator is the sum of the maximum between the forecast and the measured hourly power. In particular, this definition is consistent with a graphical representation of the error, where the numerator corresponds to the yellow area shown in Figures 9 and 10 and the denominator is the sum of the gray and yellow areas highlighted in the same figures. With reference to the above-mentioned days, while the two $NMAE_\%$ values are nearly the same, the $EMAE_\%$ is 11% in the first case and 40% in the second case, and it never exceeds 100%.

As with the daily $NMAE_{\%,d}$, in this study we also introduced the mean value of all the $EMAE_{\%,d}$, which are referred to the d-th day, calculated over the whole period. Therefore, $\overline{EMAE_\%}$ is the mean of all the daily $EMAE_{\%,d}$ for a given data-set:

$$\overline{EMAE_\%} = \frac{1}{D} \sum_{d=1}^{D} EMAE_{\%,d}. \qquad (9)$$

4. Case Study

Experimental data for this study were taken from the laboratory SolarTechLab [37] located in Milano, Italy (coordinates: 45°30'10.588" N; 9°9'23.677" E). In 2014, the DC output power of a single PV module with the following characteristics was recorded:

- PV technology: Silicon mono crystalline,
- Rated power (Net capacity of the PV module): 245 Wp,
- Azimuth: $-6°30'$ (assuming 0° as south direction and counting clockwise),
- Solar panel tilt angle (β): 30°,

The monitoring activity of the PV system parameters lasted from 8 February to 14 December 2014, but the employable data, without interruptions and discontinuities, amount to 216 days. These 24-hourly samples were used as the database for the forecasting methods comparison.

The PV module was linked to the electric grid by a micro-inverter ABB MICRO-0.25-I- OUTD [38], guaranteeing the optimization of the production. Its operating parameters—DC power included—were transmitted to a workstation for storage using a ZigBee protocol wireless connection, in real-time. An important issue that arises is how to avoid missing values and outliers. A suitable pre-processing procedure, which has already been developed and described in detail in [39], is applied here.

The weather forecasts employed were delivered by a weather service each day at 11 a.m. of the day before the forecasted one, for the exact location of the PV plant. The historical hourly database of these parameters was used to train the network and includes the following parameters:

- T_{amb} ambient temperature (°C),
- GHI global horizontal irradiance (W/m^2),
- $GPOA$ global irradiance on the plane of the array (W/m^2),
- W_s wind speed (m/s),
- W_d wind direction (°),
- P pressure (hPa),
- R precipitation (mm),
- C_c cloud cover (%),
- C_t cloud type (Low/Medium/High).

In addition to these parameters, in order to train the PHANN method, the local time LT (hh:mm) of the day and the Clear Sky Radiation model $CSRM$ (W/m^2) were also provided. These are the eleven inputs of the ANN. Regarding the specific settings of the ANN, exception made for the training database composition (as presented in Section 2), they were selected on the basis of a sensitivity analysis, as outlined in a previous study [23]. The ANN settings adopted in this study were:

- neurons in the input layer: 11,
- neurons in the first hidden layer: 11,
- neurons in the second hidden layer: 5,
- neurons in the output layer: 1,
- training algorithm: Levenberg–Marquardt,
- activation function: sigmoid,
- number of trials in the ensemble forecast: 40.

The share of the data included in the training and in the validation steps have been adjusted by means of another sensitivity analysis. Independently of how many days were employed in the training, the database was divided into two groups containing different amounts of data. Thereafter, they were provided first to train the network and the remaining data for the validation. Finally, the ensemble forecast was performed. This procedure was followed several times, progressively increasing the number of days employed in the training-process. The above-mentioned performance indexes over the whole year were calculated, and according to the different shares adopted between training and validation, the results are plotted in Figures 11 and 12. The results depicted here refer to the training method C1, and the reason for this choice will be explained later in Section 5. As can be seen, the best results are always guaranteed by adopting 90% of data for the training and the remaining 10% for the validation (the blue rhomboidal curve). However, the zoom in the top-right corner of Figure 11 shows that, for the largest amount of data (210 days), also 80% of data for the training and 20% for the validation (the purple dotted curve) provided similar results to the previously described curve. The same $\overline{NMAE}_\%$ trends were obtained in Figure 12, where the trend of $\overline{EMAE}_\%$ is shown as a function of the data-set size and the shares of training and validation set.

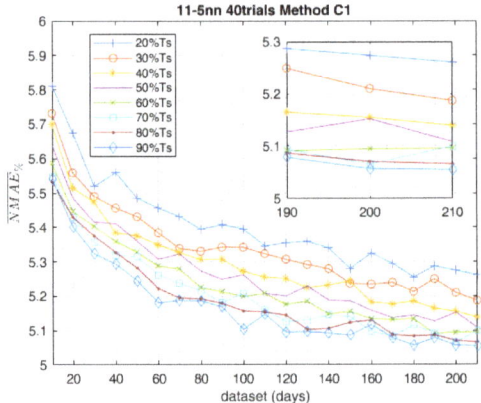

Figure 11. $\overline{NMAE}_\%$ as a function of the dataset size.

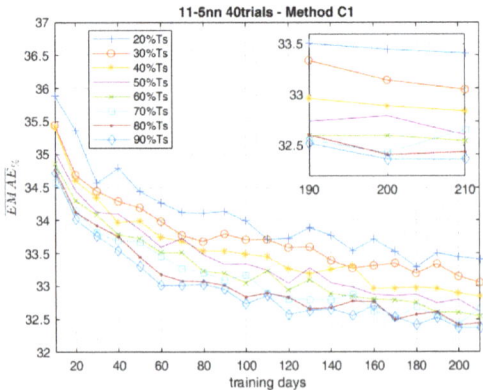

Figure 12. $\overline{EMAE}_\%$ as a function of the dataset size.

The same analysis is performed for the training Method A* by comparing the results of $\overline{NMAE}_\%$ in Figure 13 and the new error definition Equation (9) shown in Figure 14.

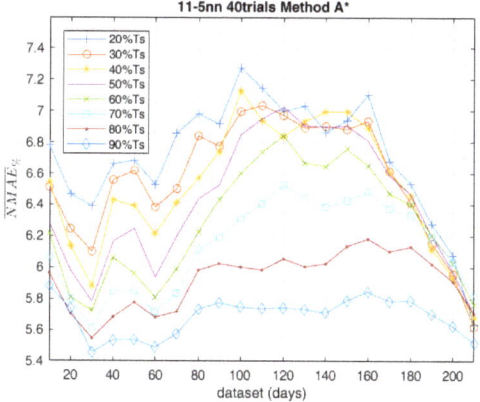

Figure 13. $\overline{NMAE}_\%$ as a function of the dataset size.

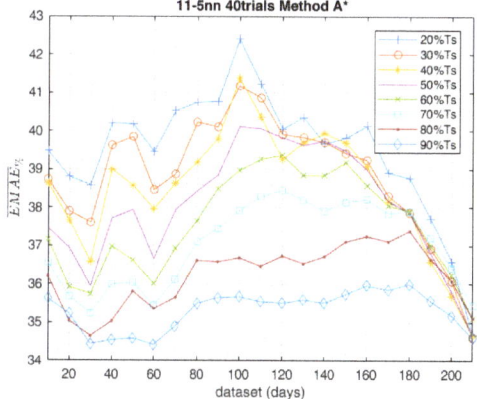

Figure 14. $\overline{EMAE}_\%$ as a function of the dataset size.

5. Results

The study carried on so far aimed to compare different methods in the data-set composition employed for the training of the ANN, highlighting the most effective ones. The obtained results of the day-ahead forecasts were analysed by the indexes shown in Section 3 and led to the following results. The graph in Figure 15 shows the trend of the $\overline{NMAE}_\%$ calculated for the methods in the training-set composition, according to increasing data-set sizes. The best training method, which globally performed better with all the data-sets considered, was undoubtedly C1. Instead, in the short-range training, with only 10 days available in the data-set, method C2 scored the worst result with $\overline{NMAE}_\%$ equal to 6.079. In accordance with the increasing data-sets method, C2 aligned with C1 above 90–130 days. The same trends of the other evaluation indexes are equally shown in Figures 16–18 and confirm the same results. From this perspective, method C2 scored the worst result, with $\overline{EMAE}_\%$ equal to 36.51. According to the $\overline{NMAE}_\%$ shown in Figure 15, methods B1 and B2 generally performed pretty much the same.

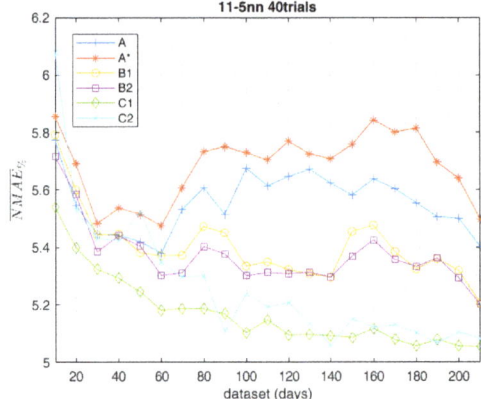

Figure 15. $\overline{NMAE}_\%$ as a function of the dataset size.

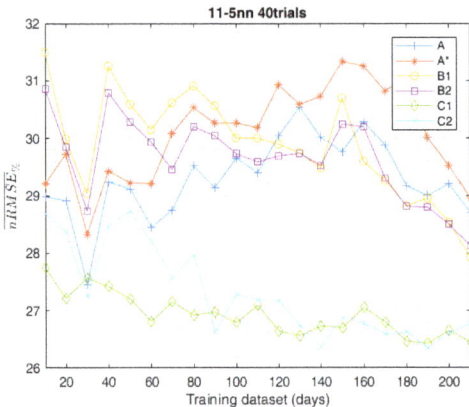

Figure 16. $\overline{nRMSE}_\%$ as a function of the dataset size.

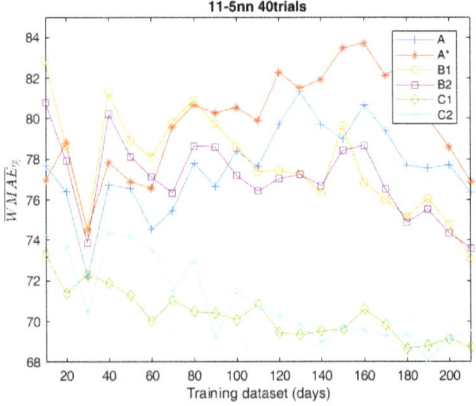

Figure 17. $\overline{WMAE}_\%$ as a function of the dataset size.

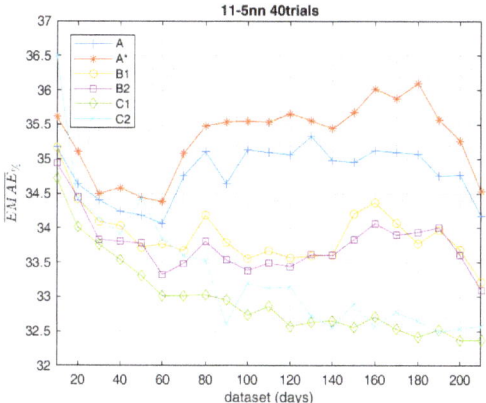

Figure 18. $\overline{EMAE}_\%$ as a function of the dataset size.

As a general comment on the reported results, it can be stated that method A is best suited when the availability of historical data is limited (e.g., newly deployed PV plant), while method C1 appears to be most effective in the case of a greater availability of data (e.g., at least one year of power measurements from the considered PV facility). Generally speaking, ensembles composed of independent trials are most effective. The performance of methods B1 and B2 was halfway between A and C, and their effectiveness in the case of newly deployed PV plants became significant after a minimum period of measurement data accumulation (above 60 days).

6. Conclusions

This paper has presented a specific study aimed to analyze the effect of different approaches in the composition of a training data-set for the day-ahead forecasting of PV power production. In particular, the authors proposed different procedures to set-up the training and validation data-sets for the ANN used in physical hybrid method to perform the power forecast in view of the electricity market. The here-outlined approaches can be adopted to set-up data-sets based on either historical data retrieved from an existing PV plant or on incremental data measurements in a newly deployed PV facility. In particular, the influence of different data-set compositions on the forecast outcome has been inspected by increasing the training dataset size and by varying the training and validation shares, in order to assess the most effective training method of this machine learning approach, based on commonly used and newly-defined performance indexes for the prediction error. The reported results have been validated over a 1-year time range of experimentally measured data from a real PV power plant, considering a comparison of various error measures and showing the best approach for the different cases of either newly deployed or already existing PV facilities.

Author Contributions: In this research activity, all of the authors were involved in the data analysis and preprocessing phase, the simulation, the results analysis and discussion, and the manuscript's preparation. All of the authors have approved the submitted manuscript. All the authors equally contributed to the writing of the paper.

Conflicts of Interest: The authors declare no conflict of interest.

References

1. Pelland, S.; Remund, J.; Kleissl, J.; Oozeki, T.; De Brabandere, K. Photovoltaic and solar forecasting: State of the art. *IEA PVPS Task* **2013**, *14*, 1–36.
2. Paulescu, M.; Paulescu, E.; Gravila, P.; Badescu, V. *Weather Modeling and Forecasting of PV Systems Operation*; Springer Science & Business Media: Berlin, Germany, 2012.

3. Raza, M.Q.; Nadarajah, M.; Ekanayake, C. On recent advances in PV output power forecast. *Sol. Energy* **2016**, *136*, 125–144.
4. Faranda, R.S.; Hafezi, H.; Leva, S.; Mussetta, M.; Ogliari, E. The Optimum PV Plant for a Given Solar DC/AC Converter. *Energies* **2015**, *8*, 4853–4870.
5. Dolara, A.; Lazaroiu, G.C.; Leva, S.; Manzolini, G.; Votta, L. Snail Trails and Cell Microcrack Impact on PV Module Maximum Power and Energy Production. *IEEE J. Photovolt.* **2016**, *6*, 1269–1277.
6. Omar, M.; Dolara, A.; Magistrati, G.; Mussetta, M.; Ogliari, E.; Viola, F. Day-ahead forecasting for photovoltaic power using artificial neural networks ensembles. In Proceedings of the 2016 IEEE International Conference on Renewable Energy Research and Applications (ICRERA), Birmingham, UK, 20–23 November 2016; pp. 1152–1157.
7. Cali, Ü. *Grid and Market Integration of Large-Scale Wind Farms Using Advanced Wind Power Forecasting: Technical and Energy Economic Aspects*; Erneuerbare Energien und Energieeffizienz—Renewable Energies and Energy Efficiency; Kassel University Press: Kassel, Germany, 2011.
8. Ni, Q.; Zhuang, S.; Sheng, H.; Kang, G.; Xiao, J. An ensemble prediction intervals approach for short-term PV power forecasting. *Sol. Energy* **2017**, *155*, 1072–1083.
9. Simonov, M.; Mussetta, M.; Grimaccia, F.; Leva, S.; Zich, R. Artificial intelligence forecast of PV plant production for integration in smart energy systems. *Int. Rev. Electr. Eng.* **2012**, *7*, 3454–3460.
10. Duan, Q.; Shi, L.; Hu, B.; Duan, P.; Zhang, B. Power forecasting approach of PV plant based on similar time periods and Elman neural network. In Proceedings of the 2015 Chinese Automation Congress (CAC), Wuhan, China, 27–29 November 2015; pp. 1258–1262.
11. Gardner, M.; Dorling, S. Artificial neural networks (the multilayer perceptron)—A review of applications in the atmospheric sciences. *Atmos. Environ.* **1998**, *32*, 2627–2636.
12. Nelson, M.; Illingworth, W. *A Practical Guide to Neural Nets*; Physical Sciences; Addison-Wesley: Boston, MA, USA, 1991; 316p.
13. Bose, B.K. Neural Network Applications in Power Electronics and Motor Drives—An Introduction and Perspective. *IEEE Trans. Ind. Electron.* **2007**, *54*, 14–33.
14. Ogliari, E.; Dolara, A.; Manzolini, G.; Leva, S. Physical and hybrid methods comparison for the day ahead PV output power forecast. *Renew. Energy* **2017**, *113*, 11–21.
15. Elder, J.F.; Abbott, D.W. A comparison of leading data mining tools. In Proceedings of the Fourth International Conference on Knowledge Discovery and Data Mining, New York, NY, USA, 27–31 August 1998; Volume 28.
16. Bergstra, J.; Breuleux, O.; Bastien, F.; Lamblin, P.; Pascanu, R.; Desjardins, G.; Turian, J.; Warde-Farley, D.; Bengio, Y. Theano: A CPU and GPU math compiler in Python. In Proceedings of the 9th Python in Science Conference, Austin, TX, USA, 28 June–3 July 2010; pp. 1–7.
17. Collobert, R.; Kavukcuoglu, K.; Farabet, C. Torch7: A matlab-like environment for machine learning. In Proceedings of the BigLearn, NIPS Workshop, Sierra Nevada, Spain, 16–17 December 2011; Number EPFL-CONF-192376.
18. Kalogirou, S. *Artificial Intelligence in Energy and Renewable Energy Systems*; Nova Publishers: Hauppauge, NY, USA, 2007.
19. Duffie, J.A.; Beckman, W.A. *Solar Engineering of Thermal Processes*; John Wiley & Sons: Hoboken, NJ, USA, 2013.
20. Gandelli, A.; Grimaccia, F.; Leva, S.; Mussetta, M.; Ogliari, E. Hybrid model analysis and validation for PV energy production forecasting. In Proceedings of the 2014 International Joint Conference on Neural Networks (IJCNN), Beijing, China, 6–11 July 2014; pp. 1957–1962.
21. Dolara, A.; Grimaccia, F.; Leva, S.; Mussetta, M.; Ogliari, E. A Physical Hybrid Artificial Neural Network for Short Term Forecasting of PV Plant Power Output. *Energies* **2015**, *8*, 1138–1153.
22. Rana, M.; Koprinska, I.; Agelidis, V.G. Forecasting solar power generated by grid connected PV systems using ensembles of neural networks. In Proceedings of the 2015 International Joint Conference on Neural Networks (IJCNN), Killarney, Ireland, 12–16 July 2015; pp. 1–8.
23. Grimaccia, F.; Leva, S.; Mussetta, M.; Ogliari, E. ANN Sizing Procedure for the Day-Ahead Output Power Forecast of a PV Plant. *Appl. Sci.* **2017**, *7*, 622.
24. Netsanet, S.; Zhang, J.; Zheng, D.; Hui, M. Input parameters selection and accuracy enhancement techniques in PV forecasting using Artificial Neural Network. In Proceedings of the 2016 IEEE International Conference on Power and Renewable Energy (ICPRE), Shanghai, China, 21–23 October 2016; pp. 565–569.

25. Panapakidis, I.P.; Christoforidis, G.C. A hybrid ANN/GA/ANFIS model for very short-term PV power forecasting. In Proceedings of the 2017 11th IEEE International Conference on Compatibility, Power Electronics and Power Engineering (CPE-POWERENG), Cadiz, Spain, 4–6 April 2017; pp. 412–417.
26. Tetko, I.V.; Livingstone, D.J.; Luik, A.I. Neural network studies. 1. Comparison of overfitting and overtraining. *J. Chem. Inf. Comput. Sci.* **1995**, *35*, 826–833.
27. Hansen, L.K.; Salamon, P. Neural network ensembles. *IEEE Trans. Pattern Anal. Mach. Intell.* **1990**, *12*, 993–1001.
28. Perrone, M.P. General averaging results for convex optimization. In *Proceedings of the 1993 Connectionist Models Summer School*; Psychology Press: London, UK, 1994; pp. 364–371.
29. Odom, M.D.; Sharda, R. A neural network model for bankruptcy prediction. In Proceedings of the 1990 IJCNN International Joint Conference on Neural Networks, San Diego, CA, USA, 17–21 June 1990; pp. 163–168.
30. Hagan, M.T.; Demuth, H.B.; Beale, M.H. *Neural Network Design*; Campus Publishing Service, University of Colorado Bookstore: Boulder, CO, USA, 2014; ISBN 9780971732100.
31. Chen, S.H.; Jakeman, A.J.; Norton, J.P. Artificial intelligence techniques: an introduction to their use for modelling environmental systems. *Math. Comput. Simul.* **2008**, *78*, 379–400.
32. Monteiro, C.; Fernandez-Jimenez, L.A.; Ramirez-Rosado, I.J.; Munoz-Jimenez, A.; Lara-Santillan, P.M. Short-Term Forecasting Models for Photovoltaic Plants: Analytical versus Soft-Computing Techniques. *Math. Probl. Eng.* **2013**, *2013*, 767284.
33. Ulbricht, R.; Fischer, U.; Lehner, W.; Donker, H. First Steps Towards a Systematical Optimized Strategy for Solar Energy Supply Forecasting. In Proceedings of the European Conference on Machine Learning and Principles and Practice of Knowledge Discovery in Databases (ECMLPKDD 2013), Riva del Garda, Italy, 23–27 September 2013.
34. Kleissl, J. *Solar Energy Forecasting and Resource Assessment*; Academic Press: Cambridge, MA, USA, 2013.
35. Ogliari, E.; Grimaccia, F.; Leva, S.; Mussetta, M. Hybrid Predictive Models for Accurate Forecasting in PV Systems. *Energies* **2013**, *6*, 1918–1929.
36. Wolfram, M.; Bokhari, H.; Westermann, D. Factor influence and correlation of short term demand for control reserve. In Proceedings of the 2015 IEEE Eindhoven PowerTech, Eindhoven, The Netherlands, 29 June–2 July 2015; pp. 1–5.
37. SolarTechLab Department of Energy. Available online: http://www.solartech.polimi.it/ (accessed on 30 September 2017).
38. ABB MICRO-0.25-I-OUTD. Available online: https://library.e.abb.com/public/0ac164c3b03678c085257cbd0061a446/MICRO-CDD_BCD.00373_EN.pdf (accessed on 21 January 2018).
39. Leva, S.; Dolara, A.; Grimaccia, F.; Mussetta, M.; Ogliari, E. Analysis and validation of 24 hours ahead neural network forecasting of photovoltaic output power. *Math. Comput. Simul.* **2017**, *131*, 88–100.

 © 2018 by the authors. Licensee MDPI, Basel, Switzerland. This article is an open access article distributed under the terms and conditions of the Creative Commons Attribution (CC BY) license (http://creativecommons.org/licenses/by/4.0/).

Article

Online Identification of Photovoltaic Source Parameters by Using a Genetic Algorithm

Giovanni Petrone [1,†,*], Massimiliano Luna [2], Giuseppe La Tona [2], Maria Carmela Di Piazza [2] and Giovanni Spagnuolo [1]

1. Dipartimento di Ingegneria dell'Informazione ed Elettrica e Matematica Applicata, Università degli Studi di Salerno, 84084-Fisciano, Italy; gspagnuolo@unisa.it
2. Consiglio Nazionale delle Ricerche, Istituto di Studi sui Sistemi Intelligenti per l'Automazione, 90146-Palermo, Italy; luna@pa.issia.cnr.it (M.L.); latona@pa.issia.cnr.it (G.L.T.); dipiazza@pa.issia.cnr.it (M.C.D.P.)

* Correspondence: gpetrone@unisa.it; Tel.: +39-08-996-4277
† Current address: Via Giovanni Paolo II n.132-84084 Fisciano, Salerno, Italy.

Received: 26 November 2017; Accepted: 15 December 2017; Published: 22 December 2017

Abstract: In this paper, an efficient method for the online identification of the photovoltaic single-diode model parameters is proposed. The combination of a genetic algorithm with explicit equations allows obtaining precise results without the direct measurement of short circuit current and open circuit voltage that is typically used in offline identification methods. Since the proposed method requires only voltage and current values close to the maximum power point, it can be easily integrated into any photovoltaic system, and it operates online without compromising the power production. The proposed approach has been implemented and tested on an embedded system, and it exhibits a good performance for monitoring/diagnosis applications.

Keywords: single-diode photovoltaic model; online diagnosis; genetic algorithm; embedded systems

1. Introduction

The photovoltaic (PV) single-diode model (SDM), shown in Figure 1, is widely used for describing the electrical behavior of a photovoltaic source because it is a good trade-off between model complexity and precision. Such a model is mainly adopted to reproduce the electrical I-V curve of the PV source, and in general, it operates offline with respect to the system under investigation. The SDM is also useful in online applications such as model-based maximum power point tracking (MPPT) and monitoring/diagnosis operation [1–4].

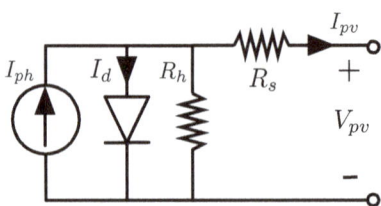

Figure 1. Equivalent circuit of the single-diode model.

The five parameters (I_{ph}, I_s, η, R_s, R_h) appearing in (1), which is the equation underlying the SDM, are usually calculated by using datasheet information, or experimental data, or a combination of them. Due to the strong nonlinearity and the involved implicit relationships, the calculation of the SDM parameters is a challenging task. In particular, this calculation involves the use of iterative algorithms that have some drawbacks: they are slow and do not guarantee the convergence to the exact solution if a good initial estimate (guess solution) is not available. Therefore, iterative methods for the SDM parameter identification are not suitable to be applied on-line, for example to support MPPT [5].

Some papers highlighted how the variation of the SDM parameters is strictly related to some degradation phenomena occurring inside the PV panel. Such degradation phenomena should be properly monitored and eventually removed to avoid significant losses in the energy production [6,7]. In this scenario, the adoption of online parameter identification procedures, working during the normal operation of the PV source, is very attractive. By comparing the identified values with the ones assumed as the reference and related to the proper operation of the source, the state of health of the PV panels can be detected.

Besides the aging issues, there are some common circumstances where the state of health of the PV panels changes suddenly as for hot-spot phenomena. In these cases, a prompt identification of the SDM parameters' variations can avoid destroying the PV panels and prevent dangerous situations like triggering a fire. In mismatched PV fields, hot-spot phenomena appear frequently, so that an on-line monitoring of the state of health is highly recommended.

For all the above reasons, several technical contributions have been proposed to perform the online parameter identification of the PV SDM. For example, in [8], a four-parameter formulation of the SDM is used to estimate the I-V curve and the maximum power point (MPP) in real time by using six pairs of voltage-current experimental points close to the MPP, whereas in [2], a software running on a personal computer (PC), which is connected to a testbed system, is used to validate the real-time implementation of such a technique.

In [9], the calculation of the SDM parameter is proposed based on explicit formulas. Specifically, the number of parameters of the SDM is reduced to four (one of the two resistances is neglected) on the basis of a suitable classification of PV panels according to their series to parallel ratio (SPR). In [10], a method that allows identifying the set of five SDM parameters by explicit formulas is given; it always keeps the fifth-order model, but in extreme conditions, it can result in unrealistic negative values of one of these two resistances. Moreover, in [11], suitable parameter translation equations, used to evaluate the SDM parameters under any environmental condition, are tested for several identification methods based on explicit formulas, and the accuracy of the translation procedure is quantitatively assessed for different case studies.

Most of the procedures described in the literature to calculate the SDM parameters require the knowledge of the short circuit current (I_{sc}) and the open circuit voltage (V_{oc}) in the actual environmental conditions. This information allows simplifying the parameters' calculation since at such points, the PV voltage and current are equal to zero; thus, some equations can be simplified. On the other hand, in order to maximize the energy production, the PV system is always controlled to work as close as possible to its maximum power point. Therefore, the measurement of I_{sc} and V_{oc} is undesirable, since in such points, the PV source delivers zero power. Moreover, the power converter used to regulate the PV source is often not properly designed to work in the short circuit or open circuit points. Therefore, additional devices and complex procedures should be introduced to perform those measurements. For these reasons, all the SDM parameter identification procedures requiring the I_{sc} and V_{oc} values are mainly effective when run offline, on the basis of previously-acquired sets of measurements.

In recent years, some authors have proposed computational intelligence-based methods for the PV source model parameter identification. The proposed methods range from genetic algorithms (GAs) to differential evolution, and examples of applications to the identification of the five-parameter SDM are given in the literature [12,13]. In the case of GAs, the core idea is to define a population of individuals where each individual is a set of parameter values and then to select the best-fitted individuals as the base for generating a new population, by minimizing an error function. The main advantage of this approach is that it does not require the use of complex equations to evaluate the model parameters.

As well as in any other numerical approaches, although the initial values could be generated randomly, providing a good guess solution significantly helps the algorithm convergence and improves the execution time. On the other hand, an inappropriate selection of the initial values will result in unacceptable parameter values or in non-convergence of the algorithm [14].

Computational intelligence-based algorithms usually require powerful computing platforms to exhibit a reasonable execution time; for this reason, up to now, the embedded implementation of such techniques has been critical, sometimes forcing designers to simplify the objective function or to discard the algorithm. However, powerful embedded platforms have recently been made available on the market, for example field programmable system-on-chip (FPSoC) and microcontrollers based on 32-bit ARM processor cores, such as the STM32 family.

As for FPSoC, a technical contribution has demonstrated that it is possible to achieve a performance comparable to that of a desktop computer when running a particle swarm optimization (PSO) algorithm [15]. As for the other platform, an STM32 microcontroller has been used for the implementation of a fixed low-order controller [16]; however, to the best of the authors' knowledge, no implementation of complex optimization algorithms on such a device family has been proposed, yet.

In this paper, a novel approach for the online SDM five-parameter identification, requiring only some measured points close to the MPP, is proposed. In particular, the values of I_{sc} and V_{oc} are properly estimated so as to avoid the loss of power deriving from their measurements. Then, the problem of the appropriate determination of the guess solution is solved by using a set of suitable explicit formulas [10]. Finally, the exact solution is obtained by running a GA. The proposed method has been implemented on a very low-cost (EUR 20.00), high-performance board, namely the NUCLEO-F429ZI, which is based on an STM32 microcontroller by STMicroelectronics (Geneva, Switzerland). The experimental results demonstrate the validity of the proposed approach.

$$I_{pv} = f(V_{pv}, I_{pv}) = I_{ph} - I_s [e^{\frac{(V_{pv} + I_{pv} R_s)}{\eta V_t}} - 1] - \frac{V_{pv} + I_{pv} R_s}{R_h} \quad (1)$$

2. The Optimized SDM Parameter Identification Method Based on Genetic Algorithm

The method proposed in this paper combines the GA, which is a common method for calculating the five parameters ($I_{ph}, I_s, \eta, R_s, R_h$) [12,17], with some explicit equations that are also used to calculate the SDM parameters in a direct way [10,11], i.e., without requiring iterative algorithms. Both methods, when applied independently, require the knowledge of the current and voltage values in the MPP and the values of I_{sc} and V_{oc} for the actual environmental condition. In the proposed solution, the measurements of I_{sc} and V_{oc} are replaced by their estimated values to obtain an approximated solution for the five parameters ($I_{ph}, I_s, \eta, R_s, R_h$). The latter is used as a guess solution in the GA algorithm and also used to constrain the GA research space in a proper way, thus allowing a fast convergence towards the optimal solution. In order to catch the right information about the I-V curvature around the MPP, some additional current and voltage values close to MPP must be measured and used in the GA fitness function to assure a precise evaluation of the SDM parameters.

2.1. Genetic Algorithm Basic Function Description

Although many advanced genetic algorithm tools are available [18–20], a basic version of GA has been selected for implementing the proposed method. This choice has been made because the main objective of this paper is to implement the technique on a low-cost digital platform, thus suitably for the online operation. The GA code has been developed in C/C++ starting from the free-download version available in [21] and distributed under the GNU Lesser General Public License license.

The GA starts by randomly generating the individuals of the initial population. The number of individuals (N) is the population size. Each individual represents a solution of the problem to be solved, and the elements composing the individuals are called genes. For the SDM parameter estimation problem, each individual is composed of five genes representing the values of ($I_{ph}, I_s, \eta, R_s, R_h$); hence, the individual is a vector of five elements. Differently from [21], in the proposed approach, the five genes of one individual of the initial population are initialized with the guess solution, which is computed as discussed in Section 2.4.

The GA evolves by modifying the population emulating the biological evolution; in fact, the new individuals are obtained by means of the following processes:

- Selector function: the individuals that survive and reproduce are selected by evaluating the cumulative fitness function.
- Crossover function: the individuals created by the Selector function can swap genes with another individual of the population (i.e., the other parent); therefore, these children inherit genes from both parents. The percentage of individuals created with this function is determined by parameter P_C.
- Mutation function: the individuals created by the Selector function can be subject to a random mutation of their genes. The percentage of individuals created with this function is determined by parameter P_M.
- Elite function: the individuals with the best fitness function in the current population are preserved in the next generation. The number of preserved individuals is specified using parameter N_E.

The Crossover, Mutation and Elitefunctions have been implemented as shown in [21]. On the other hand, the Selector function of [21] has been modified so as to speed up the execution on the embedded platform as much as possible, as discussed in Section 4.

On the basis of the values of P_C, P_M and N_E, the individuals move differently in the research space from one generation to another. The GA makes the population evolve until the maximum number of generations (N_g) is reached. The individual with the best fitness in the last generation will be the optimal solution.

2.2. Genetic Algorithm Fitness Function Calculation

The GA fitness function is evaluated by accounting for the deviation from the desired goals. A first error term is the root mean square error (RMSE) of the fitted I-V curve, given M experimental test points. Since the I-V curvature changes significantly around the MPP, the M points must be selected so as to include the MPP. For each test point $\mathbf{P_i} = [\mathbf{V_i}, \mathbf{I_i}]$, the measured current $\mathbf{I_i}$ must be compared with the current that satisfies the implicit and transcendental SDM equation (1) for $V_{pv} = \mathbf{V_i}$. To this aim, the explicit version of (1) can be obtained using the Lambert-W function:

$$I_{pv,i} = -\frac{\eta \cdot V_t}{R_s} \cdot W(\theta_I) + \frac{I_{ph} + I_s - V_{pv,i}/R_h}{1 + R_s/R_h} \quad (2)$$

with:

$$\theta_I = \frac{\left(\frac{R_h \cdot R_s}{R_h + R_s}\right) \cdot I_s \cdot e^{\frac{R_h \cdot R_s \cdot (I_{ph}+I_s) + R_h \cdot V_{pv,i}}{\eta \cdot V_t \cdot (R_h + R_s)}}}{\eta \cdot V_t} \quad (3)$$

In [10,22], many details and useful references about the Lambert-W function can be found. The numerical calculation of the Lambert-W function has been implemented in C language as shown in [23].

Once $I_{pv,i}$ is known for i in $\{1, 2, ..., M\}$, the RMSE of the fitting curve can be computed as:

$$\textbf{RMSE} = \sqrt{\frac{1}{M}\sum_{i=1}^{M}(I_{pv,i} - \mathbf{I_i})^2} \quad (4)$$

Moreover, to constrain the P-V curve to have its maximum in the MPP, the error on the derivative of the power in the MPP is also calculated, as shown in [24], thus improving the convergence and precision of the genetic algorithm [12]:

$$E_{MPP} = \left.\frac{dP}{dV}\right|_{MPP} = \frac{-\frac{1}{R_h} - \frac{I_s}{\eta \cdot V_t} \cdot e^{\frac{V_{MPP} + I_{MPP} \cdot R_s}{\eta V_t}}}{1 + \frac{R_s}{R_h} + \frac{R_s \cdot I_s}{\eta \cdot V_t} \cdot e^{\frac{V_{MPP} + I_{MPP} \cdot R_s}{\eta V_t}}} \cdot V_{MPP} + I_{MPP} \quad (5)$$

To account for both types of errors, which are independent, they must be combined in quadrature. Furthermore, since the considered genetic algorithm maximizes the objective function, the latter has been made equal to the reciprocal of the overall error:

$$\text{Fitness} = \frac{1}{\sqrt{E_{MPP}^2 + RMSE^2}} \quad (6)$$

It is worth noting that in [12,17], the fitness function is calculated by selecting test points from I_{sc} to V_{oc}; in this paper, instead, all the points are concentrated close to the MPP. Figure 2 shows the difference and highlights that the new approach allows reducing the power loss significantly because the system is controlled to operate not too far from the MPP. It is also evident that the distribution of test points must be large enough to easily catch the I-V curvature; thus, the choice of the number and position of test points around the MPP must be made as a trade-off between power loss reduction and precision in the SDM parameters' identification. This aspect is described in Section 3.1, where some experimental cases are proposed.

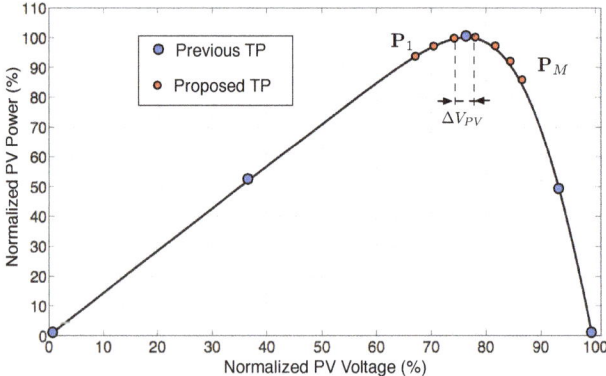

Figure 2. Test points' selection (TP) for calculating the fitness function

2.3. GA Boundary Constraint Definition

In general, the genes assigned to the individuals of a GA population are constrained in ranges depending on the problem to be solved. As explained in [12] and related references, for each SDM parameter, the bounds have been selected by considering the physical constraints, applied to the equivalent electrical circuit of the photovoltaic source shown in Figure 1, as well as typical values found in the literature where experimental tests have been performed to put into evidence the variations of those parameters for the different PV technologies and environmental conditions.

The obtained bounds are shown in Table 1. It is worth noting that, if no further information is available, the GA research space is only confined by such bounds.

Table 1. Single-diode model (SDM) parameters physical bounds.

Parameter	I_{ph} (A)	I_s (A)	η	R_s (Ω)	R_h (Ω)
Lower Bound	0	10^{-10}	0.5	0.01	10
Upper Bound	$1.5\, I_{sc,max}$	10^{-2}	5	10	100,000

The nonlinear equation representing the I-V photovoltaic curve is strongly sensitive to the parameters' variation; considering that the number of parameter combinations is high, it is very difficult to find the optimal solution when the research space is very large and the multimodality of the objective function increases the probability to be trapped in a local optimum. However, since the operating conditions and the PV technology significantly affect the SDM parameters, this information can be properly exploited to reduce the research domain with respect to that expressed by the physical bounds, thus improving the robustness of the GA to converge towards the global optimum and consequently to enhance the performance of the on-line SDM parameters' identification method. This concept is qualitatively explained with the help of Figure 3 for a bi-dimensional case (i.e., for two generic parameters P1 and P2).

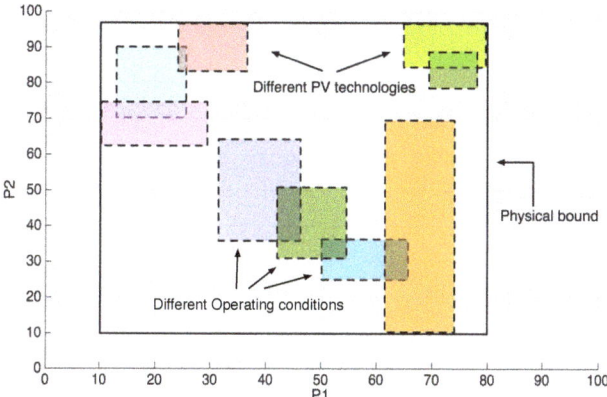

Figure 3. The difference of physical boundary conditions (continuous box) and real boundary conditions (dashed boxes).

The approach used in this paper adopts a guess solution not only to initialize the genes of one individual of the population, but also to preliminarily detect the region of the physical research domain where it is highly probable to find the best solution. Then, the GA evolves by searching the optimal solution only in this region. In particular, every time the SDM parameter identification method is activated, the guess solution $\mathbf{P_{guess}} = [I_{phg}, I_{sg}, \eta_g, R_{sg}, R_{hg}]$ is preliminarily calculated, and the new boundary conditions of the GA research space are computed as shown in Table 2.

The real bounds for the parameters having large variations (I_s, R_s, R_h) are at least one order of magnitude higher and lower with respect to the guess solution values. In this way, a wide enough range is provided to account for any error in the guess solution, which is computed using approximated explicit equations, as discussed in the next section.

Table 2. SDM parameters' real boundary conditions.

Parameter	I_{ph} (A)	I_s (A)	η	R_s (Ω)	R_h (Ω)
Lower Bound	$0.9 \cdot I_{phg}$	$0.01 \cdot I_{sg}$	max $(0.5, \eta_g - 1)$	max $(0.01, 0.1 \cdot R_{sg})$	max $(10, 0.1 \cdot R_{hg})$
Upper Bound	$1.1 \cdot I_{phg}$	$100 \cdot I_{sg}$	min $(5, \eta_g + 1)$	min $(10, 10 \cdot R_{sg})$	min $(10^5, 100 \cdot R_{hg})$

2.4. Guess Solution Calculation

The guess solution is obtained by means of the explicit equations shown in the following and is used as an approximated solution for the SDM parameters. More details are given in [11,25] and the related references.

$$I_{phg} \simeq I_{sc} \tag{7}$$

$$I_{sg} = C_0 \cdot T^3 e^{\left(-\frac{E_g}{kT}\right)} \tag{8}$$

$$\eta_g = \frac{V_{oc}}{V_t \cdot \ln\left(\frac{I_{phg}}{I_{sg}} + 1\right)} \tag{9}$$

$$R_{sg} = \frac{x\eta_g V_t - V_{MPP}}{I_{MPP}} \tag{10}$$

$$R_{hg} = \frac{x\eta_g V_t}{I_{phg} - I_{MPP} - I_{sg} \cdot (e^x - 1)} \tag{11}$$

where $V_t = \frac{kT}{q}$ is the thermal voltage of the PV junction, E_g is the material bandgap and C_0 is the temperature coefficient. The latter quantity is computed using the following equations:

$$C_0 = \frac{I_{sc0} \cdot e^{\gamma_0}}{T_0^3}; \quad \gamma_0 = -\frac{V_{oc0}}{\alpha_v - \frac{V_{oc0}}{T_0}} \left(\frac{\alpha_I}{I_{sc0}} - \frac{3}{T_0} - \frac{E_{g0}}{kT_0^2}\right) + \frac{E_{g0}}{kT_0} \tag{12}$$

The subscript "0" stands for a reference condition, which usually corresponds to standard test conditions (STC).

The auxiliary variable x is calculated by using the Lambert-W function again:

$$x = W\left[\frac{V_{MPP}\left(2I_{MPP} - I_{phg}\right) e^{\frac{V_{MPP}(V_{MPP} - 2\eta_g V_t)}{\eta_g^2 V_t^2}}}{\eta_g I_{sg} V_t}\right] + 2\frac{V_{MPP}}{\eta_g V_t} - \frac{V_{MPP}^2}{\eta_g^2 V_t^2} \tag{13}$$

As the previous equations show, datasheet information concerning the operation in standard test conditions (STC) and the thermal coefficients α_I and α_V is needed to calculate the SDM parameters. Moreover, the values of I_{sc}, V_{oc} V_{MPP}, I_{MPP} and the PV cell temperature T at the current environmental condition must also be provided.

It is worth noting that I_{sc} and V_{oc} appear only in (7) and (9), respectively. In the proposed method, to avoid the measure of I_{sc} and V_{oc} in the actual operating conditions, the following approximation will be used for silicon-based PV panels:

$$\begin{aligned} I_{sc} = \beta_I \cdot I_{MPP} & \quad \beta_I \in [1.05 \div 1.20] \\ V_{oc} = \beta_V \cdot V_{MPP} & \quad \beta_V \in [1.10 \div 1.35] \end{aligned} \tag{14}$$

The ranges of β_I and β_V have been selected as suggested in [26,27]. As a reference, average values for the silicon-based PV panels studied in [5,9,11] are $\beta_I = 1.10784$ and $\beta_V = 1.29541$. On the other hand, the dye-sensitized solar cells (DSSCs) and polymer PV modules (PPM) studied in [28] exhibit slightly higher values: $\beta_I = 1.13158$, $\beta_V = 1.48414$ for DSSCs and $\beta_I = 1.25053$, $\beta_V = 1.49043$ for PPM. Hence, the ranges of β_I and β_V should be slightly increased for non-silicon PV panels.

3. Validation of the GA-Based SDM Parameter Identification Method

This section presents the results obtained by compiling and running on a desktop PC (Intel i5-3470 quad-core processor, running at 3.2 GHz) the code written in C language to implement the GA and to calculate the guess solution. The aim is to test the code, to tune the genetic algorithm parameters and to validate the approach. A further section is dedicated to the details about the embedded system implementation.

The procedure has been applied to the experimental data of a Sunowe Solar SF125x125-72-m(l) PV panel. The related datasheet parameters are given in Table 3.

Table 3. Sunowe Solar SF125x125-72-m(l) 180-W PV panel.

Parameter	Value	Parameter	Value
I_{sc}	5.32 A	V_{oc}	44.8 V
I_{MPP}	5.03 A	V_{MPP}	35.8 V
α_I	0.04 %/°C	α_V	−0.35 %/°C
NOCT = 45 ± 2 °C.			

Figure 4 shows the experimental data used to test the proposed SDM parameter identification procedure. The white curves are the measured I-V curves, acquired in different irradiance and temperature conditions. The black points are the test points selected to be used by the proposed SDM parameter identification procedure. Specifically, the case related to Test #1 is discussed hereinafter, whereas the cases related to Tests #2 and #3 will be explained in Section 4.

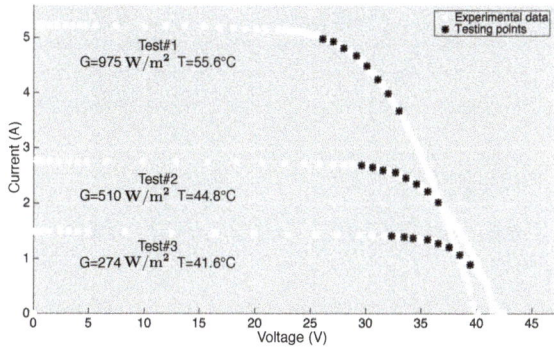

Figure 4. Experimental I-V curves of the Sunowe Solar SF125x125-72-m(l) PV panel, in different environmental conditions.

The module temperature (T) is measured by a sensor placed at the backside of the PV panel. This quantity is used in the identification procedure for calculating the guess solution. In the absence of such a sensor, T can be estimated by using the ambient temperature, as described in [25].

The guess solution for the first experimental case has been calculated by using Equations (7)–(11) with $\beta_I = 1.2$ and $\beta_V = 1.35$. The corresponding parameters are reported in Table 4.

It is worth noting that different guess solutions can be calculated by changing the values of β_I and β_V within their respective ranges. The choice of the best guess solution can be made by evaluating the corresponding fitness with (6). For the case of Table 4, the fitness value is reported in the last column.

Table 4. Guess solutions for the SDM parameters in Test #1.

Parameter	I_{ph} (A)	I_s (A)	η	R_s (Ω)	R_h (Ω)	Fitness
Test #1	5.61	5.58 × 10^{-8}	1.05	0.833	51.5	1.56

In order to appreciate the benefit of using a restricted research space for the SDM parameter identification, the genetic algorithm has been launched twice: using the physical bounds of Table 1 and using the real boundary conditions of Table 2, calculated on the basis of the guess solution reported in Table 4. When the real boundary conditions are used, the guess solution is also included as an individual of the initial population; thus, a further improvement is obtained in the GA convergence.

The parameters of the GA have been set as reported in Table 5; they have been selected on the basis of the fitness function behavior. In order to show the different behavior, Figure 5 shows the trend of the best individual's fitness with respect to the number of generations for the two above-mentioned scenarios: using only the physical bounds (Figure 5a) and using the real boundary constraints and the corresponding guess solution (Figure 5b). The figure is related to the case Test #1, but it is also representative of the other two cases. The improvement is evident in terms of the higher fitness value and faster convergence.

Table 5. Genetic algorithm parameters.

Parameter	Value	Parameter	Value
Population size	$N = 150$	Elite individuals	$N_E = 1$
Number of generations	$N_g = 2500$	Crossover percentage	$P_C = 80\%$
Number of testing points	$M = 8$	Mutation percentage	$P_M = 40\%$

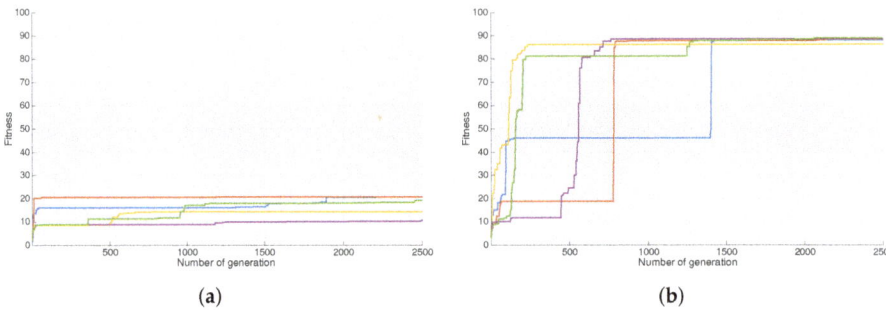

(a)   (b)

Figure 5. Best fitness value vs. the number of generations for different runs of the genetic algorithm. (a) With physical bounds; (b) with real boundary conditions and the guess solution.

Once the GA has returned the best solution, the latter has been used to reconstruct the I-V and P-V curves of the PV panel. Then, the reconstructed curves have been compared with the ones obtained using the guess solution and with the experimental data. The plots are shown in Figure 6, which refers to the case Test #1. As expected, the guess solution does not fit the experimental data in the regions far from the MPP since approximated values of I_{sc} and V_{oc} have been used. Instead, the best solution returned by the GA allows reproducing the correct I-V curvature since the information coming from the M experimental test points has been exploited.

Finally, in Table 6, the best GA solution is compared with the guess solution: R_h and I_s are the parameters that have been affected by the main variations after the refinement of the guess solution performed by the GA.

Table 6. SDM parameters' comparison for Test #1.

Parameter	I_{ph} (A)	I_s (A)	η	R_s (Ω)	R_h (Ω)	Fitness
Guess solution (A)	5.61	5.58×10^{-8}	1.05	0.833	51.5	1.56
Best GA solution (B)	5.19	4.45×10^{-6}	1.40	0.922	3953	80.59
Variation (B/A)	0.92	79.7	1.33	1.11	76.7	51.67

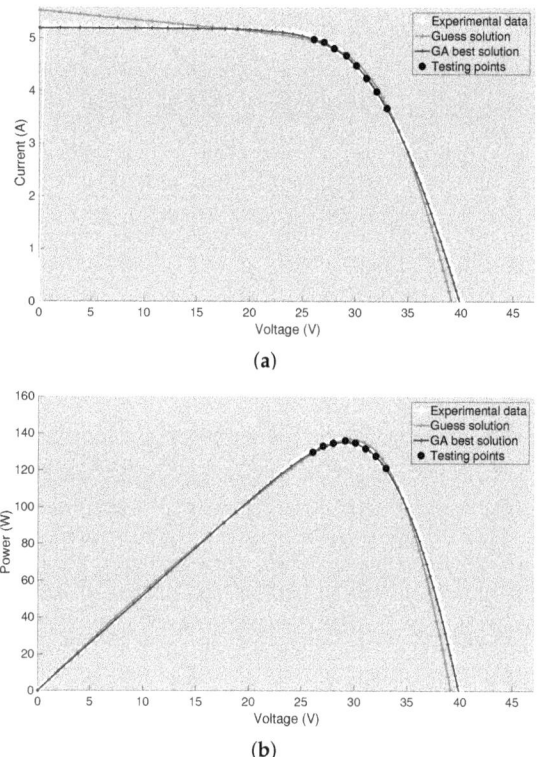

Figure 6. Comparison of the I-V and P-V photovoltaic curves for Test #1. (**a**) Photovoltaic I-V curves; (**b**) photovoltaic P-V curves.

3.1. Test Point Selection

The maximum variation of the I-V curvature occurs in proximity to the MPP, while the I-V curve is almost linear near I_{sc} and V_{oc}. Thus, the number and positions of the test points must be chosen as a trade-off between the minimum distance from the MPP (reduced power loss) and the right I-V curvature identification (high precision in the SDM parameter calculation). As a reference, the authors of [7] focused on the identification of R_s, and they suggested selecting test points up to 60% ÷ 75% of I_{MPP}. On the other hand, a 15% voltage reduction to the left of V_{MPP} is usually enough to enter the other nearly linear portion of the I-V curve. Since the PV source is usually controlled by regulating the PV voltage, it is very simple to acquire the PV voltage and current by increasing (or decreasing) the control voltage reference in a step-by-step manner, thus moving with high precision around the MPP. For the case under study, the test points are equally spaced by $\Delta V = 1V$. To cover the desired range, N = 8 points are enough, and 4 V is the maximum distance from the MPP.

As highlighted in the previous sections, the main benefit of the proposed approach is the reduction of the power loss during the measurement of the test points. For the case under study, in the worst case, the power delivered during the acquisition of the test points is only 11% less than the one delivered in the MPP, as shown in Figure 7.

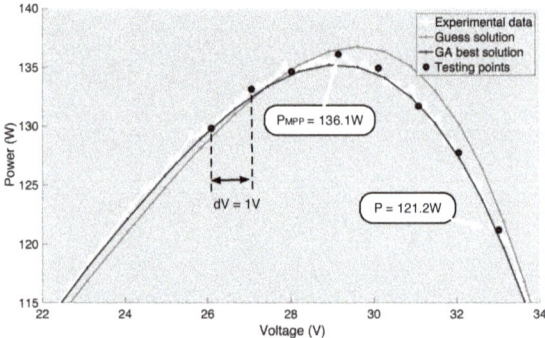

Figure 7. Zoom of the P-V photovoltaic curves for Test #1.

4. Performance Evaluation of the Embedded GA-Based Method

The platform chosen for the embedded implementation of the proposed GA-based parameter identification algorithm is the NUCLEO-F429ZI board by STMicroelectronics. Despite being a very low-cost board (EUR 20.00), it encompasses an STM32F429ZI microcontroller that is based on a high-performance Cortex-M4 32-bit RISC (Reduced Instruction Set Computing) core by ARM (Cambridge, United Kingdom), operating at up to 180 MHz and capable of up to 225 DMIPS. Such a microcontroller is equipped with 2 MB of Flash memory and 256 kB of SRAM. It features a rich variety of internal peripherals, among which three 12-bit, analog to digital converters (ADCs) and two expansion connectors that allow using a wide choice of specialized shields. Suitable signal conditioning circuits have been set up to extend the ADC voltage range; the obtained measurement system exhibits a 100-V voltage range and a 10-A current range. A virtual serial port over a USB connection has been used to print debug messages and statistics. The need to generate the 48-MHz clock for the USB port has imposed a maximum working frequency of 168 MHz for the microprocessor. Nonetheless, a frequency of 180 MHz can be used if the USB communication is not required.

After the validation on a desktop PC described in Section 3, the functions written in C language to implement the GA and to calculate the guess solution have been integrated into a previously implemented digital controller for the switching converter of the PV panel. Figure 8 shows the functional scheme and the flowchart of the enhanced digital controller, which encompasses a voltage controller, a perturb and observe (P&O) MPPT algorithm and the proposed online parameter identification algorithm. With reference to Figure 8b, the white blocks represent the typical flowchart of the perturb and observe algorithm, whereas the gray blocks allow performing the I-V scan and activating the online parameter identification procedure. In Figure 8b, Tp is the period of the MPPT algorithm, thus it is in the order of magnitude of milliseconds. The on-line identification method can be managed by an interrupt service routine activated by an additional timer having a periodicity of hours or minutes, depending on the objective of the monitoring procedure.

The whole C project has been compiled with no relevant modifications for the STM32 microprocessor and experimentally tested. When the parameter identification procedure is triggered, for example by a timer or on demand, the MPPT algorithm is temporary disabled, and the PV operating point is driven to the left of the MPP at $V_{pv} \simeq V_{MPP} - \frac{M}{2} \Delta V$. Then, the I-V curve is scanned and the voltage and current values corresponding to the M operating points close to the MPP are stored in memory. Subsequently, the guess solution and the real bounds are calculated. Once the GA has been configured, it is launched, and the MPPT algorithm is reactivated. It is worth noting that the time spent calculating the best solution of the SDM parameters does not affect the control dynamics of the DC/DC converter since this task runs concurrently with the MPPT algorithm until a new request of the SDM parameter identification is triggered.

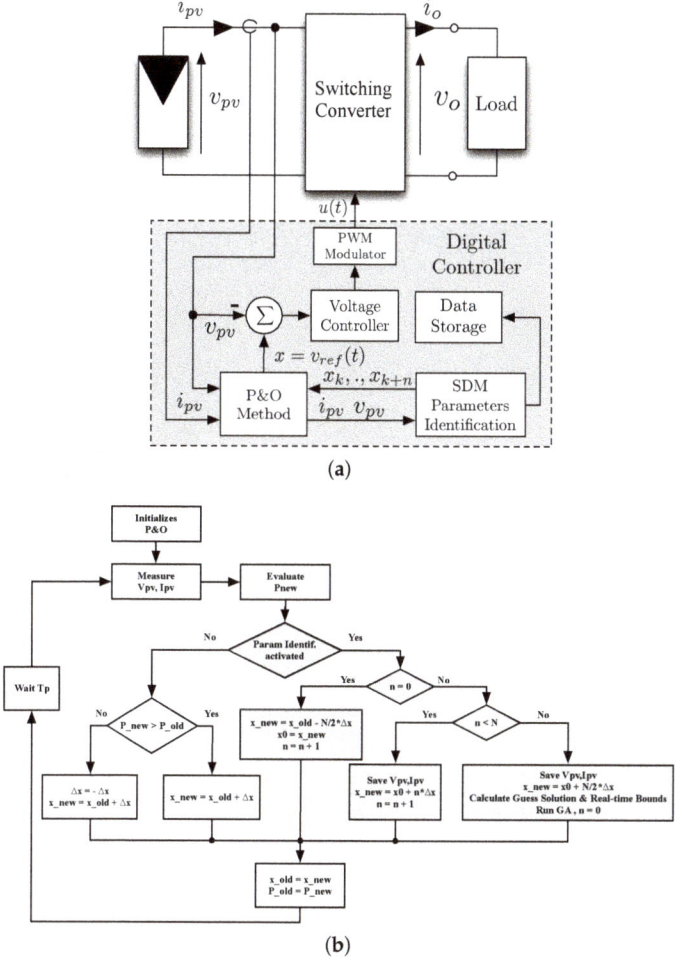

Figure 8. Photovoltaic system with the MPPT and online parameter identification technique. (a) Functional scheme of the proposed digital controller; (b) flowchart of the perturb and observe algorithm integrated with the parameter identification procedure.

The experimental validation has been performed on the same Sunowe Solar PV panel of Section 3 in the other two operating conditions (Tests #2 and #3). The white curves of Figure 4 are the whole I-V curves measured offline, whereas the black points are the M test points acquired online by the embedded system.

The guess solutions and the best solutions returned by the GA for the two experimental cases are shown in Table 7. Once again, these solutions have been used to reconstruct the I-V and P-V curves of the PV panel in the two considered operating conditions. Then, the reconstructed curves have been compared with the ones obtained using the guess solution and with the experimental data referred to each test case. The plots are shown in Figures 9 and 10 and show that the best solutions returned by the GA allows a correct I-V curve reproduction.

Table 7. SDM parameters comparison for Tests #2 and #3.

Parameter	Parameter	I_{ph} (A)	I_s (A)	η	R_s (Ω)	R_h (Ω)	Fitness
Test #2	Guess solution (A)	3.07	3.56×10^{-9}	1.09	1.9	102	2.23
Test #2	Best GA solution (B)	2.78	3.56×10^{-7}	1.36	1.37	4405	85.33
Test #2	Variation (B/A)	0.90	100	1.25	0.72	43.2	38.26
Test #3	Guess solution (C)	1.62	5.96×10^{-10}	1.13	4.1	206	5.24
Test #3	Best GA solution (D)	1.46	5.30×10^{-9}	1.13	0.876	740.2	25.96
Test #3	Variation (D/C)	0.90	8.89	1.004	0.214	3.59	4.95

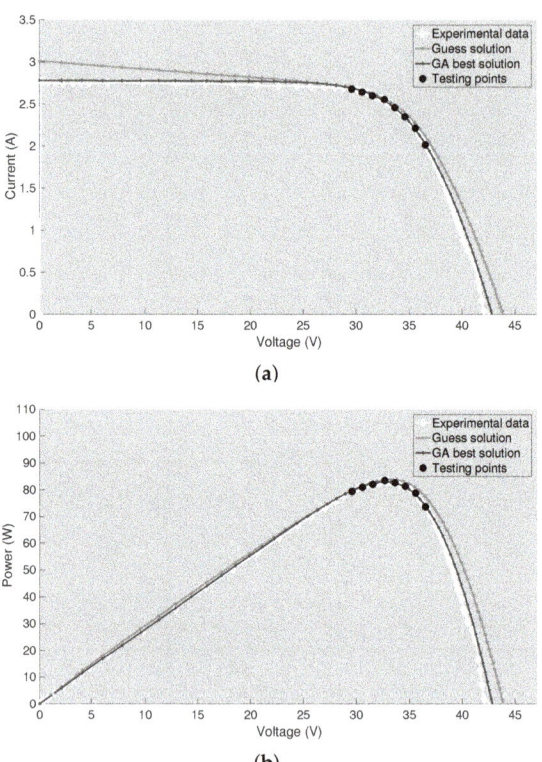

Figure 9. Comparison of the I-V and P-V photovoltaic curves for Test #2. (**a**) Photovoltaic I-V curves; (**b**) photovoltaic P-V curves.

As for the numerical results returned by the GA, there is no difference in comparison with the execution on the desktop PC because the C compiler for the STM32 microcontroller supports the double data type. The execution time on the microcontroller, instead, is very different from that on the desktop PC. Aiming to reduce it as much as possible, the Selector function of the GA available in [21] has been suitably modified. In particular, to pick up the individuals that survive and reproduce, the original function used a linear search over a vector holding the values of the cumulative fitness function. The worst-case computational complexity of the linear search is $O(n)$. Instead, in the proposed approach, a binary search algorithm is used, which has a worst-case computational complexity of $O(log(n))$. Table 8 reports the average execution time of the GA in different conditions and shows that an appreciable speed gain has been obtained. Overall, the obtained execution time of the GA using the binary search on the STM32 microcontroller is under 10 min, thus more than adequate for on-line monitoring/diagnosis applications.

Table 8. Execution time of the genetic algorithm.

Platform	Search Algorithm	Time	Variation
Desktop PC	linear search	1430 ms	-
Desktop PC	binary search	1381 ms	−3.4%
STM32	linear search	609,251 ms	-
STM32	binary search	573,903 ms	−5.8%

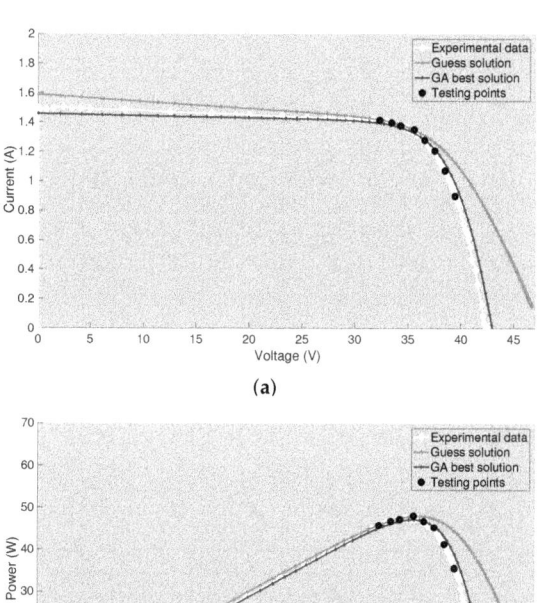

Figure 10. Comparison of the I-V and P-V photovoltaic curves for Test #3. (**a**) Photovoltaic I-V curves; (**b**) photovoltaic P-V curves.

5. Conclusions

An SDM parameters' identification method has been discussed in this paper. The proposed approach combines the simplicity of an explicit method, used for calculating an approximated solution, with the exploring capability of the genetic algorithm; the latter is adopted for finding the best solution in a properly selected research space. Accurate results have been achieved without using the direct measurement of short circuit current and open circuit voltage that are typically used in other parameter identification methods. Since the proposed method requires only voltage and current values close to the maximum power point, it can be easily integrated into any photovoltaic system, and it operates online without compromising the power production. The combined method has been implemented and tested on a very low-cost STMicroelectronics NUCLEO-F429ZI, exhibiting good performance and confirming the potential of such a kind of embedded system for achieving the online monitoring/diagnosis of PV plants.

Acknowledgments: This work was supported by the fund: "Fondo di Ateneo per la Ricerca di Base" (FARB projects) of the University of Salerno.

Author Contributions: Giovanni Petrone and Massimiliano Luna conceived the methodology and developed the simulations, they also contributed to write the paper, to perform the experiments and to analyze data; Giuseppe La Tona worked to develop the software and to the perform the tests; Maria Carmela Di Piazza and Giovanni Spagnuolo contributed to develop the methodology, to write the paper and validate the data of the experimental tests.

Conflicts of Interest: The authors declare no conflict of interest.

References

1. Manganiello, P.; Ricco, M.; Petrone, G.; Monmasson, E.; Spagnuolo, G. Optimization of Perturbative PV MPPT Methods Through Online System Identification. *IEEE Trans. Ind. Electron.* **2014**, *61*, 6812–6821.
2. Blanes, J.M.; Toledo, F.J.; Montero, S.; Garrigós, A. In-Site Real-Time Photovoltaic I-V Curves and Maximum Power Point Estimator. *IEEE Trans. Power Electron.* **2013**, *28*, 1234–1240.
3. Wang, W.; Liu, A.C.F.; Chung, H.S.H.; Lau, R.W.H.; Zhang, J.; Lo, A.W.L. Fault Diagnosis of Photovoltaic Panels Using Dynamic Current-Voltage Characteristics. *IEEE Trans. Power Electron.* **2016**, *31*, 1588–1599.
4. Batzelis, E.I.; Kampitsis, G.E.; Papathanassiou, S.A. Power Reserves Control for PV Systems With Real-Time MPP Estimation via Curve Fitting. *IEEE Trans. Sustain. Energy* **2017**, *8*, 1269–1280.
5. Cannizzaro, S.; Di Piazza, M.C.; Luna, M.; Vitale, G. Generalized classification of PV modules by simplified single-diode models. In Proceedings of the 2014 IEEE 23rd International Symposium on Industrial Electronics (ISIE), Istanbul, Turkey, 1–4 June 2014; pp. 2266–2273.
6. Bastidas-Rodriguez, J.D.; Franco, E.; Petrone, G.; Ramos-Paja, C.A.; Spagnuolo, G. Model-Based Degradation Analysis of Photovoltaic Modules Through Series Resistance Estimation. *IEEE Trans. Ind. Electron.* **2015**, *62*, 7256–7265.
7. Sera, D.; Mathe, L.; Kerekes, T.; Teodorescu, R.; Rodriguez, P. A low-disturbance diagnostic function integrated in the PV arrays' MPPT algorithm. In Proceedings of the IECON 2011—37th Annual Conference of the IEEE Industrial Electronics Society, Melbourne, Australia, 7–10 November 2011; pp. 2456–2460.
8. Toledo, F.; Blanes, J.M.; Garrigós, A.; Martínez, J.A. Analytical resolution of the electrical four-parameters model of a photovoltaic module using small perturbation around the operating point. *Renew. Energy* **2012**, *43*, 83–89.
9. Cannizzaro, S.; Di Piazza, M.C.; Luna, M.; Vitale, G. PVID: An interactive Matlab application for parameter identification of complete and simplified single-diode PV models. In Proceedings of the 2014 IEEE 15th Workshop on Control and Modeling for Power Electronics (COMPEL), Santander, Spain, 22–25 June 2014; pp. 1–7.
10. Accarino, J.; Petrone, G.; Ramos-Paja, C.A.; Spagnuolo, G. Symbolic algebra for the calculation of the series and parallel resistances in PV module model. In Proceedings of the 2013 International Conference on Clean Electrical Power (ICCEP), Alghero, Italy, 11–13 June 2013; pp. 62–66.
11. Di Piazza, M.C.; Luna, M.; Petrone, G.; Spagnuolo, G. Translation of the Single-Diode PV Model Parameters Identified by Using Explicit Formulas. *IEEE J. Photovolt.* **2017**, *7*, 1009–1016.
12. Bastidas-Rodriguez, J.; Petrone, G.; Ramos-Paja, C.; Spagnuolo, G. A genetic algorithm for identifying the single diode model parameters of a photovoltaic panel. *Math. Comput. Simul.* **2017**, *131*, 38–54.
13. Zagrouba, M.; Sellami, A.; Bouaïcha, M.; Ksouri, M. Identification of PV solar cells and modules parameters using the genetic algorithms: Application to maximum power extraction. *Sol. Energy* **2010**, *84*, 860–866.
14. Di Piazza, M.C.; Vitale, G. *Photovoltaic Sources: Modelling and Emulation*; Green Energy and Technology; Springer: London, UK, 2013.
15. Molanes, R.F.; Garaj, M.; Tang, W.; Rodriguez-Andina, J.J.; Farina, J.; Tsang, K.F.; Man, K.F. Implementation of Particle Swarm Optimization in FPSoC devices. In Proceedings of the 2017 IEEE 26th International Symposium on Industrial Electronics (ISIE), Edinburgh, UK, 19–21 June 2017; pp. 1274–1279.

16. Ben Hariz, M.; Bouani, F.; Ksouri, M. Implementation of a fixed low order controller on STM32 microcontroller. In Proceedings of the International Conference on Control, Engineering & Information Technology (CEIT'14), Sousse, Tunisia, 22–25 Mar 2014; pp. 244–252.
17. Ismail, M.S.; Mahlia, T.M.I.; Moghavvemi, M. Characterization of PV panel and global optimization of its model parameters using genetic algorithm. *Energy Convers. Manag.* **2013**, *73*, 10–25.
18. Global Optimization Toolbox. Available online: https://www.mathworks.com/products/global-optimization.html (accessed on 28 November 2017).
19. GPdotNET–Artificial Intelligence Tool. Available online: https://gpdotnet.codeplex.com (accessed on 28 November 2017).
20. A Java-Based Evolutionary Computation Research System. Available online: http://cs.gmu.edu/~eclab/projects/ecj/ (accessed on 28 November 2017).
21. Genetic Algorithm C++ Code. Available online: http://people.sc.fsu.edu/~jburkardt/cpp_src/simple_ga/simple_ga.html (accessed on 28 November 2017).
22. Petrone, G.; Spagnuolo, G.; Vitelli, M. Analytical model of mismatched photovoltaic fields by means of Lambert W-function. *Sol. Energy Mater. Sol. Cells* **2007**, *91*, 1652–1657.
23. Keith Briggs: Software—C and Python Codes. Available online: http://keithbriggs.info/software.html (accessed on 19 December 2017).
24. Lo Brano, V.; Ciulla, G. An efficient analytical approach for obtaining a five parameters model of photovoltaic modules using only reference data. *Appl. Energy* **2013**, *111*, 894–903.
25. Petrone, G.; Ramos-Paja, C.A.; Spagnuolo, G. *Photovoltaic Sources Modeling*; John Wiley & Sons: Chichester, UK, 2017.
26. Noguchi, T.; Togashi, S.; Nakamoto, R. Short-current pulse-based maximum-power-point tracking method for multiple photovoltaic-and-converter module system. *IEEE Trans. Ind. Electron.* **2002**, *49*, 217–223.
27. Orozco-Gutierrez, M.L.; Spagnuolo, G.; Ramirez-Scarpetta, J.M.; Petrone, G.; Ramos-Paja, C.A. Optimized Configuration of Mismatched Photovoltaic Arrays. *IEEE J. Photovolt.* **2016**, *6*, 1210–1220.
28. Cannizzaro, S.; Di Piazza, M.C.; Luna, M.; Vitale, G.; Calogero, G.; Citro, I. Parameter Identification and Real-Time J-V Curve Reconstruction of Polymer PV and Dye-Sensitized Cells Using Non-Iterative Algorithm. In Proceedings of the 29th European PV Solar Energy Conference and Exhibition (EUPVSEC 2014), Amsterdam, The Netherlands, 22 September 2014; pp. 1548–1553.

© 2017 by the authors. Licensee MDPI, Basel, Switzerland. This article is an open access article distributed under the terms and conditions of the Creative Commons Attribution (CC BY) license (http://creativecommons.org/licenses/by/4.0/).

Article

Thermal and Performance Analysis of a Photovoltaic Module with an Integrated Energy Storage System

Manel Hammami [1,*], Simone Torretti [1], Francesco Grimaccia [2] and Gabriele Grandi [1]

[1] Department of Electrical, Electronic, and Information Engineering, University of Bologna, 40136 Bologna, Italy; simone.torretti@studio.unibo.it (S.T.); gabriele.grandi@unibo.it (G.G.)
[2] Department of Energy, Politecnico di Milano, 20156 Milano, Italy; francesco.grimaccia@polimi.it
* Correspondence: manel.hammami2@unibo.it; Tel.: +39-051-2093589

Received: 19 September 2017; Accepted: 23 October 2017; Published: 25 October 2017

Abstract: This paper is proposing and analyzing an electric energy storage system fully integrated with a photovoltaic PV module, composed by a set of lithium-iron-phosphate (LiFePO$_4$) flat batteries, which constitutes a generation-storage PV unit. The batteries were surface-mounted on the back side of the PV module, distant from the PV backsheet, without exceeding the PV frame size. An additional low-emissivity sheet was introduced to shield the batteries from the backsheet thermal irradiance. The challenge addressed in this paper is to evaluate the PV cell temperature increase, due to the reduced thermal exchanges on the back of the module, and to estimate the temperature of the batteries, verifying their thermal constraints. Two one-dimensional (1D) thermal models, numerically implemented by using the thermal library of Simulink-Matlab accounting for all the heat exchanges, are here proposed: one related to the original PV module, the other related to the portion of the area of the PV module in correspondence of the proposed energy-storage system. Convective and radiative coefficients were then calculated in relation to different configurations and ambient conditions. The model validation has been carried out considering the PV module to be at the nominal operating cell temperature (NOCT), and by specific experimental measurements with a thermographic camera. Finally, appropriate models were used to evaluate the increasing cell batteries temperature in different environmental conditions.

Keywords: photovoltaic; battery; integrated storage; PV cell temperature; thermal model; thermal image

1. Introduction

Photovoltaic energy is one of most promising among renewable energy sources. Future development of photovoltaic and, in general, of all clean and renewable technologies, is related to the possibility to use energy when needed. For this reason, energy storage represents a key component in the development of renewable energies [1–4]. Energy storage makes it possible to meet users needs during out-of-production periods. In this way, renewable energies can be really competitive with fossil sources and nuclear energies. Nowadays, electrochemical storage (i.e., battery) represents the most used and reliable technology to store electrical energy. Battery energy storage systems (BESS) represent one of the most promising and flexible solutions for storage [5,6]. An effective and flexible implementation strategy of storage in energy systems is to join the storage elements to the renewable generation units. For this purpose, an electric energy storage system fully integrated within a single photovoltaic (PV) module [7], constituting a modular generation-storage PV unit, is proposed and evaluated in this paper. In this way, each PV module can be treated as a self-rechargeable battery unit (Figure 1), and the whole PV generation-storage system can be simply built by properly combining the PV units (i.e., series/parallel/connections). In particular, each PV unit would be composed of a commercial PV module, a dc/dc chopper with both maximum power point tracking

MPPT and battery charge regulator capabilities, a charge monitoring system (BMS), and a set of lithium-iron-phosphate (LiFePO$_4$) flat batteries for energy storage (typical thickness less than 1 cm), connected in series. The batteries were surface-mounted on the back side of the PV module by an aluminum-bar structure which keeps the batteries distant from the PV backsheet without exceeding the PV frame size (typical air-gap was about 2 cm). Thus, the battery disposition didn't exceed the original PV size, but allowed natural air cooling of the backsheet. An additional low-emissivity, aluminum sheet was introduced to shield the batteries from the backsheet irradiance, in consideration of the usual maximum temperature limit of the batteries (50–55 °C). This paper is aimed at evaluating the PV cell temperature increase due to the reduced thermal exchanges on the back PV side, and to estimate battery temperature in order to verify their thermal limits. A PV model based on material, environmental parameters, and electro-thermal characteristics, was developed, taking into account the energy balance. The entire solar radiation incident on PV module was converted into electrical and heat energy. Consequently, excessive heat and thermal stress can result in cell fault and/or energy losses [8]. Two one-dimensional (1D) thermal models accounting for all the heat exchanges are proposed. The first one concerns the original (commercial) PV module, the second one is related to the PV module with the proposed energy-storage system mounted on the back side. Convective and radiative coefficients were calculated in relation to different configurations and environmental conditions. The considered models were numerically implemented using the thermal library of Simulink-Matlab. The model validation were carried out by the PV module normal operating cell temperature(NOCT) given by the manufacturer, and by specific experimental measurements on the real PV module, including thermographic camera images, with and without the proposed BESS.

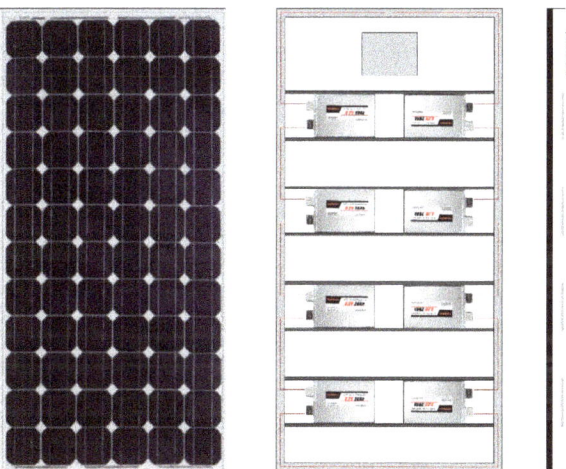

Figure 1. Rendering of the PV module with integrated batteries on the back side.

2. Thermal Model of a Single PV Module

This section deals with simple thermal models for PV modules, with and without the integrated energy storage system. The heat exchanges were theoretically analyzed and compared with reference to the two aforementioned models. The first aim of this analysis is to determine the temperature increase in PV cells in the case of the presence of the storage system, considering the different heat exchanges.

2.1. Models

For a PV module, the steady-state thermal balance can be written as [9]:

$$0 = GA - P_{pv} - Q_{tot}, \text{ being } \begin{cases} G = G_n \cdot A - G_{rif} \cdot A \\ G_{rif} = \rho \cdot G_n \\ P_{pv} = \eta \cdot G_n \cdot A \\ Q_{tot} = Q_{conv,f} + Q_{rad,f} + Q_{conv,b} + Q_{rad,b} \end{cases} \quad (1)$$

With reference to the thermal balance (Equation (1)), the heat exchanges of the commercial PV module are described in Figure 2a: G_n is the specific radiation incident on the module's surface (W/m^2); G_n must be multiplied by the area of the PV module A (m^2) in order to get the total incident power. P_{pv} (W) is the electrical power generation, proportional to the total incident power, and η is the photovoltaic conversion efficiency. G_{rif} is the total radiative power reflected from the surface of the PV module that is proportional to the reflection index ρ.

The Q terms [W] consider convective Q_{conv} and radiative Q_{rad} exchanges of the front (f) and back (b) sides of the PV module with the surrounding environment.

As for the commercial PV module, the heat exchanges of the PV module with an energy storage system are shown in Figure 2b. By introducing flat batteries on the back side of the PV module it is evident that both convective and radiative heat exchanges were limited. In particular, radiative exchange was limited due to the reflection of infrared (IR) rays by the aluminum flat plates (having extremely low emissivity). The PV module backsheetemited IR rays, proportional to its temperature (T^4). The aluminum plate acted as a thermal shield, reducing overheating of the batteries. However, the IR rays reflected from the backsheet increased the temperature of the PV module. In addition, the presence of batteries led to a restriction in the convective heat exchange, even though air circulation was still possible through the airgap (i.e., almost 2 cm).

The decrease of the radiative and conductive exchanges of the backsheet due to the batteries led to higher backsheet and PV cell temperatures compared to commercial PV modules.

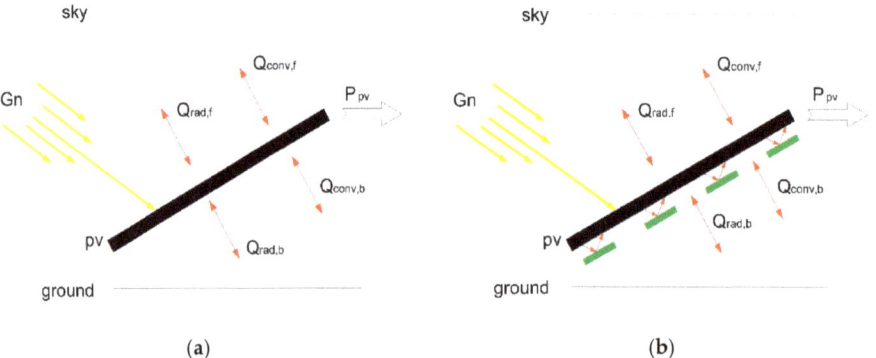

Figure 2. Heat exchanges of PV module: (**a**) without battery storage system; (**b**) with storage battery system. G_n is the specific radiation incident on the module's surface (W/m^2); P_{pv} (W) is the electrical power generation, proportional to the total incident power; η is the photovoltaic conversion efficiency; G_{rif} is the total radiative power reflected from the surface of the PV module that is proportional to the reflection index ρ.

In the thermal models under consideration, the following assumptions have been made:

- One-dimensional (1D) thermal models.
- Isothermal surface was approximated as a flux node, so edge effects were neglected.

- Negligible thermal capacitances [10].
- Material properties of PV module layers were constant, as shown in Table 1.
- PV cell temperature was considered uniform due to the higher value of its thermal conductivity (k) compared to any of the other layers.
- Apparent sky temperature T_s was calculated according to [11]:

$$T_s = T_a - \delta T \qquad (2)$$

where T_a is the air temperature and δT is the variation depending on atmospheric conditions.
- Ground temperature (on the back side of the PV module) was equal to air temperature ($T_g = T_a$).
- Convective heat exchange coefficients were evaluated with empirical formulations and it was assumed that wind flowed around the module, both front and back sides [12].
- Reflection (ρ), transmission (τ), and absorption (α) coefficients were independent from temperature.
- Internal reflection phenomena between the layers of PV module were neglected.
- Emissivity of surfaces were independent from temperature and wavelength, values are given in Table 2.
- According to the radiative heat transfer, the view factor was assumed to be unity. With this approximation, the front surface of PV module saw only the sky, whereas the back surface saw only the ground. Radiative heat exchange coefficients were simplified by this assumption.
- All surfaces had the same area A of the PV module: $A = 1.31\ m^2$ in the case study.
- Batteries did not produce any heat flow while charging or discharging (battery losses were neglected).

Table 1. Thickness (s), thermal conductivity (k), and optical coefficients (ρ, τ, α) of the PV module layers.

Layer's Material	s [mm]	k [W/(m·K)]	ρ	τ	α
Glass	4.0	1.8	0.1	0.88	0.02
Ethylene Vinyl Acetate (EVA)	0.4	0.35	-	0.97	0.03
Silicon PV cell	0.4	150	-	-	1
Backsheet	0.3	0.3	-	-	1

Table 2. Emissivity coefficients.

	ε
PV module front surface (glass) (f)	0.91
PV module back surface (backsheet) (b)	0.85
Sky	0.91
Ground	0.94
Polished aluminum plates	0.04

2.2. Thermal Balance Equations

A PV module contains a number of layers from the front to the back side as follows: glass (g), EVA (e'), silicon PV cell, another EVA layer (e''), and backsheet (b) [13]. The batteries were mounted on the back side of the PV module by an aluminum-bar structure so as to leave an airgap ($d \approx 2$ cm) between the batteries and the PV backsheet, as depicted in Figure 3.

The thermal balance for each layer can be expressed taking into account all the thermal power exchanges such as convective (*conv*), radiative (*rad*), and conductive (*cond*) powers Q. The incident thermal power on the surface of the PV module is transmitted to the different layers (G_1, G_2, G_3, and

G_4) on the basis of absorption and transmission coefficients. The summary of these thermal balances can be expressed as:

$$\begin{cases} Q_{conv,f} + Q_{rad,f} - Q_{cond,g} + G_1 = 0 \\ Q_{cond,g} - Q_{cond,e'} + G_2 = 0 \\ Q_{cond,e'} - Q_{cond,e''} + G_3 - P_{pv} = 0 \\ Q_{cond,e''} - Q_{cond,b} + G_4 = 0 \\ -Q_{conv,b} + Q_{rad,b} + Q_{cond,b} = 0 \end{cases}, \text{ being } \begin{cases} G_1 = \alpha_f \cdot G_n \cdot A \\ G_2 = \tau_f \cdot \alpha_{EVA} \cdot G_n \cdot A \\ G_3 = \tau_f \cdot \tau_{EVA} \cdot (\alpha_c K + \alpha_{EVA}(1-K))G_n \cdot A \\ G_4 = \tau_f \cdot \tau_{EVA} \cdot \tau_{EVA} \cdot \alpha_b (1-K)G_n \end{cases} \quad (3)$$

where K represents the ratio between the area of silicon of the PV cells and the total PV module area.

In addition to the previous thermal balances (Equation (3)), in the case of the PV module with energy storage, the thickness of the batteries (*batt*) can be taken into account as:

$$Q_{conv,batt-b} + Q_{rad,batt-g} + Q_{cond,al+batt} + Q_{conv,batt-g} + Q_{rad,batt-g} = 0 \quad (4)$$

where $Q_{cond,al+batt}$ is the conductive heat transfer between the battery layer and aluminum sheet.

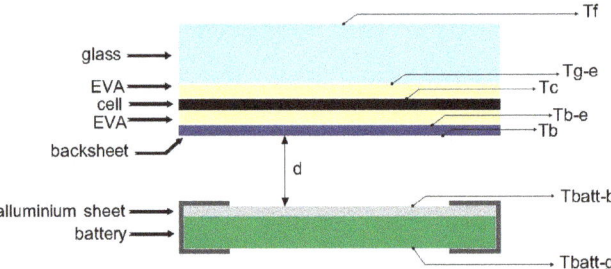

Figure 3. Battery and PV module layers. T is temperature; b is backsheet; e is ethylene vinyl acetate (EVA); f is front side; g is ground and batt is battery.

2.3. Convective and Radiative Coefficients

Convective and radiative coefficients are calculated according to the geometry of the PV module and ambient conditions such as wind speed, air temperature, and ground temperature.

The convective heat exchange is given by:

$$Q_{conv} = h_{conv} \cdot A \cdot \Delta T, \text{ being } h_{conv} = \frac{Nu \cdot k}{L} \quad (5)$$

being h_{conv} [W/(m^2 K)] the convective coefficient, A the exchange area (m^2), ΔT the difference between two surfaces at different temperatures, L the characteristic length of the geometry of the PV module (area/perimeter), k the thermal conductivity of the fluid for a reference temperature [W/(m K)], and Nu the dimensionless Nusselt number. In addition to the Nusselt number, the following numbers are helpful in order to determine the heat transfer:

$$Pr = \frac{v}{\alpha}, \; Gr = \frac{g \cdot \beta \cdot \Delta T \cdot L}{v}, \; Ra = Gr \cdot Pr = \frac{g \cdot \beta \cdot \Delta T \cdot L^3}{v \cdot \alpha}, \; Re = \frac{u \cdot L}{v} \quad (6)$$

where the Prandtl number (*Pr*) gives the information about the type of fluid. It also provides information about the thickness of the thermal and hydrodynamic boundary layer. Reynolds number (*Re*) gives information about whether the flow is inertial or viscous force dominant, in order to determine if the flow is laminar or turbulent. The Grashoff number (*Gr*) is used in the correlation of heat and mass transfer due to the thermally induced natural convection of a solid surface immersed in a fluid.

The Rayleigh number (Ra) is defined as the product of the Grashof number and the Prandtl number, and it describes the relationship between momentum diffusivity and thermal diffusivity. Parameters of interest in (Equation (6)) are: dynamic viscosity v (m^2/s); thermal diffusivity α (m^2/s); gravity acceleration g (m/s^2); isobaric compression ratio β (1/K); temperature difference ΔT, and wind speed u (m/s). All these thermal properties are evaluated at the reference temperature $T_{ref} = (T_a + T_w)/2$, where T_a is the air temperature (i.e., the fluid that is surrounding the PV module) and T_w is temperature of the surface under consideration (K).

The transition from laminar to turbulent flow was determined by the critical values (cr): $Ra_{cr} = 10^9$, $Re_{cr} = 10^5$, $Gr_{cr} = 10^9$. Measurements were taken when the wind speed was not too high, so Reynolds number was usually $Re < Re_{cr}$. In this way Nusselt number for forced convection is empirically calculated as [14]:

$$Nu_{forced} = 0.664 Re^{0.5} Pr^{1/3} \tag{7}$$

In natural convection, empirical Churchill and Chu formulas are used to calculate the convective coefficient for the PV module front and back surfaces. The Churchill–Bernstein equation is valid for a wide range of Reynolds and Prandtl numbers, and it can be used for any object of cylindrical geometry in which boundary layers develop freely, without constraints imposed by other surfaces [15], leading to:

$$Nu_{natural} = \left(0.825 + \frac{0.387 Ra^{1/6}}{\left[1 + \left(\frac{0.492}{Pr}\right)^{9/16}\right]^{8/27}}\right)^2 \tag{8}$$

Note that the Nusselt number (Equation (8)) does not change in a forced convection field, whereas it does change from the front to back sides of the PV module in natural convection. For the back side, Nu has to hold on gravity force, so it is important to correct the Rayleight number with a coefficient $g \cos\beta$ [13]. The g is the acceleration due to gravity, and the $\cos\beta$ is the tilt angle of PV module. The buoyancy force on the tilt plane has to be divided into parallel and perpendicular components. On the front side these components do not interfere with the fluid and both contribute to the natural heat exchange. However, on the back side, the speed of the fluid interferes with the surface. The ratio between the Grashoff number (Gr) and the second power of the Reynolds number (Re^2) is calculated and, according to Gr/Re^2, evaluates the nature of convection: natural, forced or mixed [14]: $Gr/Re^2 \sim 1$ means mixed convection; $Gr/Re^2 \gg 1$ natural convection; $Gr/Re^2 \ll 1$ forced convection. In the case that convection is mixed and the flows of natural and forced convection are in opposition, the largest Nu is considered to be the proper value. In the case that the flows in mixed convection are not in opposition, Nu is considered as:

$$Nu = \sqrt[3]{(Nu_{forced})^3 + (Nu_{natural})^3} \tag{9}$$

In order to estimate the PV cell temperature, it is important to correctly state the physical problem of the heat exchanges between the two plates, and in particular the convection exchanges. The temperature of the air standing around the planes is taken into account in this paper, and for this reason natural convection is considered. Figure 4 shows the geometric arrangement used to calculate the natural convection between two plane plates (i.e., backsheet and battery), where T_b is the temperature of the backsheet, and T_{batt-b} is the temperature of the battery, L_1 and L_2 are the dimensions of the first and the second plane, respectively.

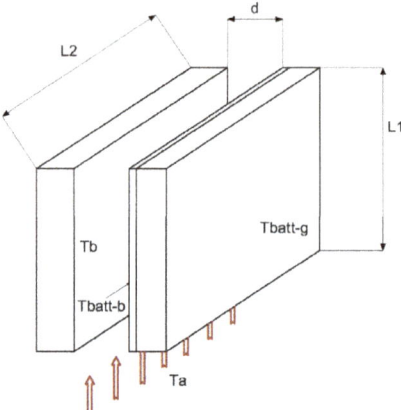

Figure 4. Natural convection between plane plates (i.e., backsheet and battery). d is the distance between the PV unit and the battery; T_b is the temperature of backsheet; and T_{batt-b} is the temperature of the battery; Ta is the temperature of the air; L_1 and L_2 are the dimensions of the first and the second plane, respectively.

According to [14], the Ra number is given by:

$$Ra_{planes} = \frac{g \cdot \beta \cdot (T_w{}^* - T_a)}{\nu \cdot \alpha} \cdot \left(\frac{d}{L_1}\right) \tag{10}$$

where $T_w{}^*$ (K) is the average temperature of the plates. The Nu number can be evaluated considering the case of two opened planes surrounded by air, as follows [14]:

$$Nu = \left[(Nu_{fd})^{-1.9} + (0.62 \cdot Ra^{1/4})^{-1.9}\right]^{-1/1.9} \tag{11}$$

$$Nu_{fd} = \frac{(4 \cdot T^{*2} - 7 \cdot T^* + 4) \cdot Ra_{planes}}{90 \cdot (1 + T^*)^2}, \text{ being } T^* = \frac{(T_b - T_a)}{(T_{batt-b} - T_a)} \tag{12}$$

By these assumptions, air temperature T_a, backsheet temperature T_b, and battery temperature T_{batt-b} are taken into account. It should be considered that the critical value of the Ra number has never reached ($Ra < Ra_{cr} = 10^5$), which justifies the validity of this approach.

The radiative heat exchange (Q_{rad}) is simplified by considering parallel planar bodies with the same area (A), and the radiative heat coefficient (h_{rad}) is calculated with reference to the unity view factor (i.e., $F_{12} = 1$), leading to:

$$Q_{rad} = h_{rad} \cdot A \cdot \left(T_1^4 - T_2^4\right), \text{ being } h_{rad} = \frac{\sigma}{\frac{1-\varepsilon_1}{\varepsilon_1} + \frac{1}{F_{12}} + \frac{1-\varepsilon_2}{\varepsilon_2}} = \frac{\sigma}{\frac{1-\varepsilon_1}{\varepsilon_1} + 1 + \frac{1-\varepsilon_2}{\varepsilon_2}} \tag{13}$$

where T_1 and T_2 are the surface temperatures (K), ε_1 and ε_2 are the corresponding emissivity coefficients, and σ is the Stephan-Boltzmann constant [5.67×10^{-8} W/(m^2 K^4)]. Equation (13) is used to evaluate all the radiative exchanges, i.e., sky-PV front side, backsheet-ground, backsheet-aluminum plate, battery-ground, considering the corresponding areas.

3. Models Results and Measurements

3.1. Simulink Thermal Models

A one-dimensional thermal model was implemented by the thermal library of Simulink-Matlab. By properly setting the boundary conditions, it was possible to calculate the temperature between layers and surfaces, with particular reference to PV cells and batteries. The schemes of the thermal models under consideration are depicted in Figure 5, with reference to the commercial PV module (Figure 5a) and the modified PV module with battery energy storage (Figure 5b). This last model makes reference to the area corresponding to the batteries.

Temperatures T_a, T_s and T_g were set by thermal blocks as ideal temperature sources. Thermal fluxes through the different layers (G_1, G_2, G_3 and G_4) were modeled by ideal heat flux sources. All the heat exchanges were modeled according to the previous sections, properly setting the parameters. Temperatures of the different thermal nodes were detected and displayed by ideal temperature sensors.

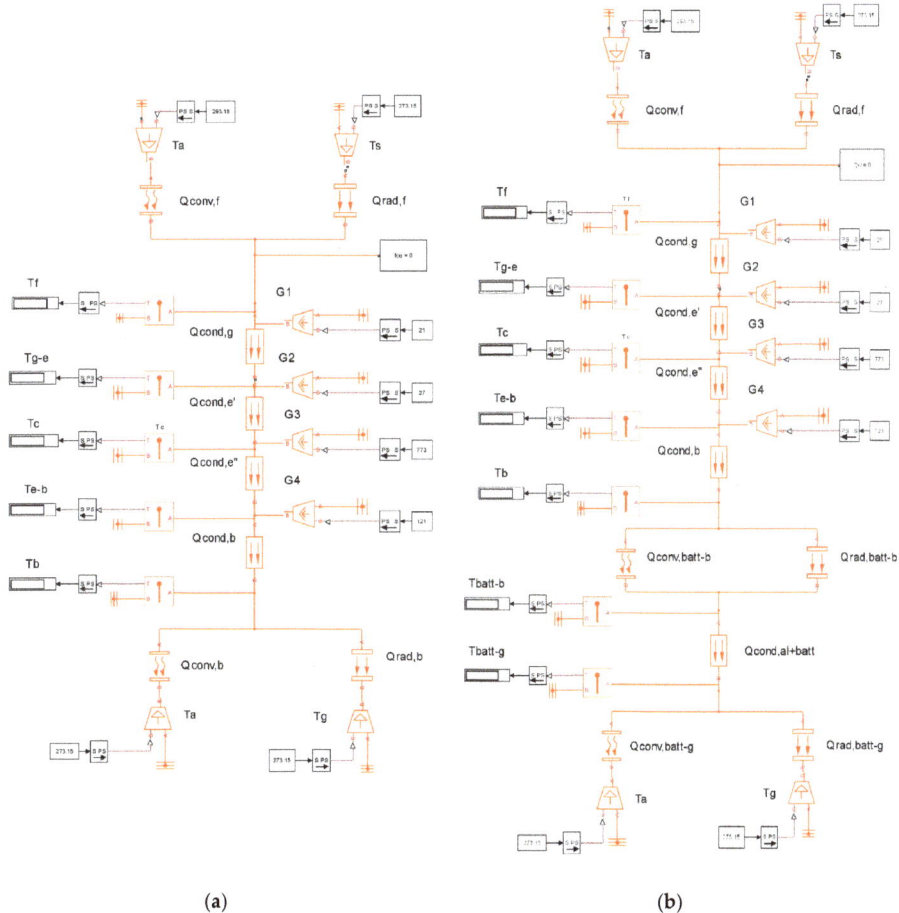

Figure 5. Simulink thermal model without (**a**) and with (**b**) battery energy storage.

3.2. Thermal Model Validation

In order to validate the proposed Simulink thermal model, the calculated temperature of the PV cell was compared with the NOCT given by the PV module manufacturer. The environment parameters defining the NOCT were: wind speed 1 m/s; air temperature 20 °C; solar radiation 800 W/m^2; tilt angle of PV module 45°. In the case study, NOCT = 46 °C (the temperature-power coefficientis $-0.52\%/°C$). The corresponding PV cell temperature obtained by the Simulink thermal model (without storage batteries) was ≈45 °C, with an acceptable matching.

Thermal models were also validated by experimental measurements on real PV module prototypes, with and without the battery storage system, using a thermographic camera, an anemometer, and a few thermometers. At first, the thermographic camera was set up with precision in consideration of the reflected temperature compensation (RTC), distance of object, relative humidity, and air temperature. After these settings were established, several thermal pictures were taken, and the temperatures were compared. Some example results are given in Figures 6 and 7. It is interesting to notice that, as expected, front surface temperature increased in the area corresponding to the back side battery areas (Figure 6). Measurements were taken in different ambient conditions and thermal pictures of both front and back sides of PV module were taken, for both the cases with and without storage batteries. The hotspots introduced in the PV module by the batteries obviously cause a slight decrease in the voltage of the corresponding PV cells (approx. $-0.5\%/°C$), and a general decrease in the PV module power and efficiency [16].

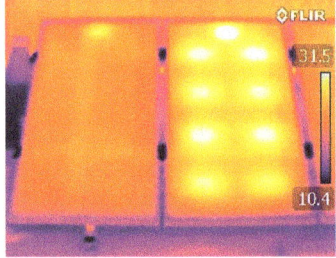

Figure 6. Thermal (**left**) and visual (**right**) images of the PV modules with and without the battery storage system for ambient conditions $T_a \cong 8\ °C$ and $G_n \cong 600\ W/m^2$.

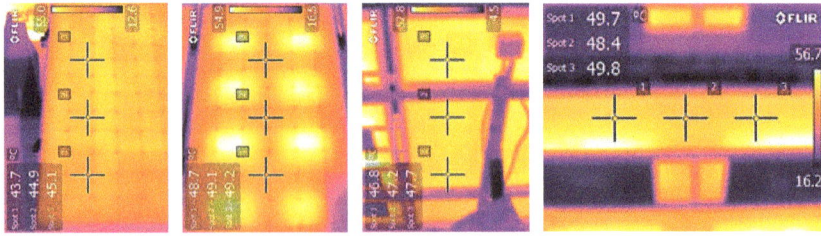

Figure 7. Thermal images of front and back sides of the PV modules, with and without the battery storage system for ambient conditions $T_a \cong 15\ °C$ and $G_n \cong 900\ W/m^2$.

Temperatures estimated by the thermographic camera (T_f, T_b, T_{batt-b}, T_{batt-g}) were properly averaged, and compared to the corresponding temperatures obtained by the thermal model. In general, the matching was satisfactory in the areas without the batteries, the difference in temperature being in the order of 1 °C. In the areas occupied by the batteries the difference was in the order of 3 °C. The difference was greater in the model with the batteries, but it was still somewhat acceptable, considering that all the edge effects were neglected in the 1D thermal model.

An additional validation, only for the model without batteries, was made comparing the PV cell temperature obtained by the Simulink thermal model to the corresponding values calculated with the formula [17] using the NOCT, where T_a and G_n were typical for the given month. In this case, again, the matching was satisfactory with the difference of temperature being within 1 °C.

Figure 8 shows the measured thermal behavior of the PV modules with and without the mounted battery storage system, and with an external flat panel completely covering the backsheet. Figure 9 shows further details about back side module thermal behavior (with and without BESS) and the mounted battery with relative layout.

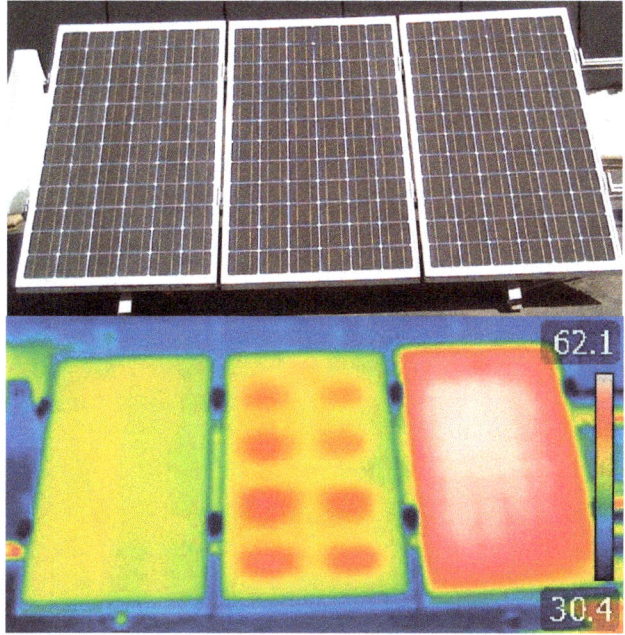

Figure 8. Visual and thermal images of the PV modules: without (**left**) and with (**center**) the battery storage system, and with covered backsheet (**right**), for ambient conditions $T_a \cong 24$ °C and $G_n \cong 800$ W/m^2.

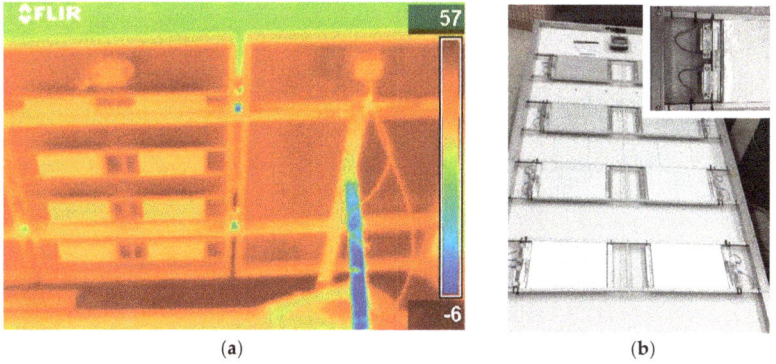

(a)   (b)

Figure 9. Back side view details of PV modules: (a) thermal image with and without battery storage system; (b) visual picture of the installed battery and layout.

4. Temperature Extrapolation

In order to determinate the over-temperature limit of PV cells due to the battery placement in the back side of the commercial PV module in different environmental conditions, the proposed thermal models were evaluated by setting different inputs (T_a and G_n) corresponding to the average temperature for each month of the year, at the solar midday. Specifically, the reference conditions are in accordance with an installation of Bologna, Italy (latitude 44°30′, tilt angle of PV modules 30°). Concurrently, the temperature of the batteries was determined in order to verify the restrictions given by the manufacturer. Thermal powers G_1, G_2, G_3 and G_4 were calculated according to Equation (3). Wind speed was set to 1 m/s for both the front to back sides of PV module (i.e., mixed convection).

Table 3 shows the resulting temperatures for each month, in the worst case scenario without electric power conversion (open circuit, $P_{pv} = 0$). In particular, for the commercial PV model (without batteries) T_f, $T_{c,pv}$, and T_b represent the front, PV cell, and back temperatures, according to Figure 3. The same temperatures are shown for the two relevant regions of the modified PV module with integrated energy storage: the area covered by the batteries, and the remaining area (not covered by batteries). In this case, the average PV cell temperature $T_{c,pv+batt}$ was calculated as the weighted average of $T_{c,pv}$ in the two regions, with respect to the size of the corresponding areas. The temperatures of the batteries were considered as well, both of the backsheet and the ground side (T_{batt-b} and T_{batt-g}), according to Figure 3. In the end, the over-temperature ΔT was calculated to illuminate the difference in PV cell temperatures between the original commercial PV module and the modified one with the integrated energy storage.

Regarding the modified PV module with integrated energy storage, the thermal exchanges related to the area not covered by batteries were treated similarly to the case of the commercial PV module without batteries. In this case, a reduced convective coefficient (decreased to 20%) was taken into account, and a unity view factor for the radiative exchange with the ground was still assumed, but a reduced radiative surface was assumed (decreased to 80%).

Table 3 shows that the thermal radiation shield (aluminum plates) causes an important increase in the PV cell temperature in the area covered by batteries, in the order of 20 °C. However, because of this shield, the over-temperature of the batteries compared to the air was limited to 5–10 °C, and the temperature of the batteries didn't exceed 40 °C, far from the limit given by the manufacturer for this type of storage element (i.e., max 50–55 °C).

Table 3 referred to the worst case scenario of open circuit operation. Table 4 shows the temperature in the case of active PV conversion ($\eta = 12\%$) in order to understand the corresponding decrease, estimated at about 4–5 °C over the year. In general, with or without PV conversion, the average PV cell temperature over the module area in the case of energy storage exceeds the corresponding temperature of the commercial PV module of 10–15 °C.

The last column of Table 4 reports the decrease (in percent) of the electric power produced by the PV module with integrated storage batteries, obtained in consideration of the over-temperature ΔT and the temperature-power coefficient ($-0.52\%/°C$) for the case study. Corresponding to Table 4, Figure 10 shows the power produced by the PV module with integrated storage batteries (in percent) compared to the power produced by the original PV module without the batteries. It should be noted that the decrease is rather noticeable, in the order of 6%, but this is representing the worst case scenario (i.e., the highest daily radiation and temperature considered for each month at solar midday on a sunny day).

Table 3. Environmental data and temperatures obtained by thermal models (open circuit, $P_{pv}=0$).

	°C	W/m²	Commercial PV Module Without Batteries (°C)				Modified PV Module with Integrated Energy Sotrage Area Covered by Batteries (°C)					Area Not Covered by Batt. (°C)			avg (°C)	(°C)
	T_a	G_n	T_f	$T_{c,pv}$	T_b	T_f	T_c	T_b	$T_{batt\text{-}b}$	$T_{batt\text{-}g}$	T_f	T_c	T_b	$T_{c,pv+batt}$	ΔT
January	2.9	741	26.6	27.7	27.2	42.6	44.6	44.5	9.3	9.3	33.9	35.4	35.1	38.9	11.2
February	5.8	869	33.4	34.8	34.1	52.0	54.3	54.2	13.3	13.3	41.7	43.4	43.0	47.5	12.8
March	12.1	994	42.7	44.3	43.6	63.6	66.2	66.1	20.3	20.3	51.5	53.5	53.0	58.3	14.0
April	17.0	1079	49.6	51.3	50.5	71.6	74.4	74.3	26.0	26.0	58.6	60.7	60.2	65.9	14.6
May	21.5	1106	54.2	55.9	55.6	76.4	79.3	79.2	30.3	30.2	63.0	65.2	64.6	70.5	14.6
June	26.3	1107	58.5	60.2	59.4	80.0	82.9	82.8	34.8	34.8	66.9	69.1	68.5	74.3	14.1
July	28.9	1107	60.7	62.4	61.6	82.0	84.9	84.8	37.3	37.3	68.9	71.1	70.5	76.3	13.9
August	28.3	1095	59.8	61.6	60.7	81.0	83.8	83.8	36.6	36.6	68.0	70.2	69.7	75.4	13.8
September	24.1	1034	54.4	56.1	55.3	74.8	77.5	77.4	32.2	32.1	62.5	64.6	64.1	69.5	13.4
October	17.0	914	44.5	45.9	45.6	63.2	65.6	65.5	24.6	24.6	52.3	54.1	53.7	58.5	12.5
November	10.0	777	33.9	35.2	34.7	50.4	52.4	52.3	16.5	16.4	41.2	42.8	42.4	46.4	11.2
December	4.8	701	26.9	28.0	27.5	41.8	43.6	43.6	10.5	10.4	33.8	35.2	34.9	38.4	10.4

Table 4. PV cell temperatures obtained by thermal models in the case of PV conversion ($P_{pv} \neq 0$, $\eta = 12\%$) and corresponding percentage decrease of electric power.

	Temperatures (°C)			Power
	$T_{c,pv}$	$T_{c,pv+batt}$	ΔT	$\Delta P\%$
January	24.3	34.1	9.8	−5.1
February	30.9	42.1	11.2	−5.8
March	40.0	52.3	12.3	−6.4
April	46.7	59.6	12.9	−6.7
May	51.4	64.3	12.9	−6.7
June	55.7	68.2	12.4	−6.5
July	58.0	70.2	12.3	−6.4
August	57.1	69.3	12.2	−6.3
September	51.8	63.6	11.8	−6.1
October	42.0	53.0	11.0	−5.7
November	31.8	41.5	9.8	−5.1
December	24.8	33.8	9.1	−4.7

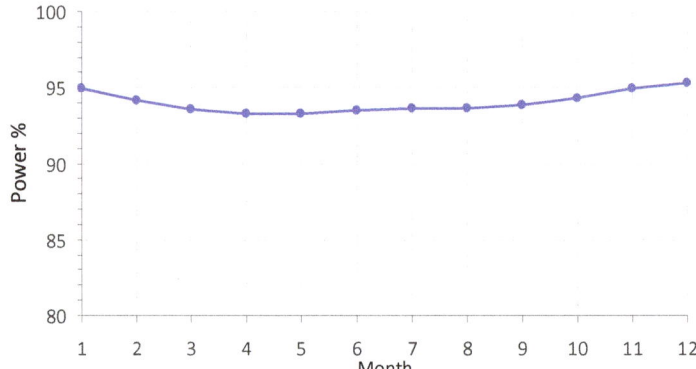

Figure 10. Estimation of the power from the PV module with integrated storage batteries in % compared to the original PV module without batteries, corresponding to Table 4.

As a further example of the application of the proposed thermal model, the daily profile of the PV cell's temperature (averaged over the PV module area) with and without integrated storage batteries is given in Figure 11. Weather conditions were assumed to be a typical sunny day in Bologna on April 15.

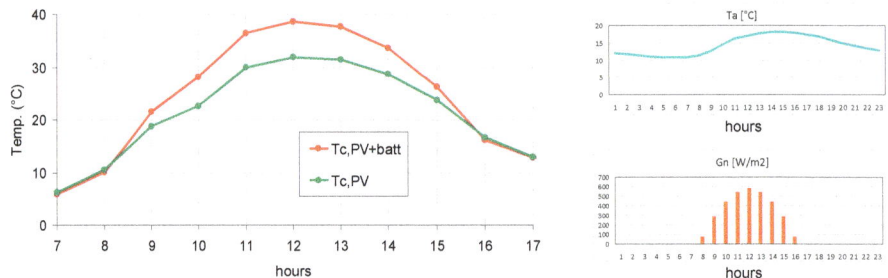

Figure 11. Average PV cells temperature profile over the day (April 15) with and without integrated storage batteries, and corresponding air temperature (T_a) and sun irradiance (G_n).

5. Conclusions

The thermal analysis of a commercial PV module with an integrated energy storage system has been carried out in this paper. Energy storage was implemented by a set of thin flat batteries placed on the back side of the module, without exceeding the original thickness of the aluminum frame. An airgap was intentionally introduced between the batteries and backsheet, for the sake of air cooling, with a thermal radiation shield to prevent overheating of the batteries. The thermal analysis was carried out by introducing a simplified 1D thermal model for the PV module, with and without the batteries.

Thermal models were numerically implemented by Simulink/Matlab, and different verification tests were carried out to validate the model results. Both PV cells and batteries temperature were extrapolated by the proposed thermal models, considering the different environmental conditions within the 12 months of the year (using midday values). In general, it has been proved that the batteries' temperature does not exceed 40 °C, which leaves a safety margin as compared to the maximum operating temperature given by the manufacturer (50–55 °C).

As expected, despite the cooling airgap, there is a remarkable PV cell temperature increase introduced by the back side batteries, estimated at 20–25 °C for the PV cells on the battery area.

Considering the average PV cell temperature over the whole PV module, this increment is reduced to 10–15 °C, representing a reasonable over-temperature in terms of the restriction of the PV conversion efficiency. A first estimation of the electric power decrease in real operating conditions is in the order of 6%, considering the worst case scenario for each month of the year (solar midday on a sunny day).

By the proposed thermal model, the analysis can be readily extended to the average PV module temperature over the whole year in order to precisely estimate the reduction in annual production of electric energy and to evaluate the overall performance of the integrated PV generation-storage system. A more detailed investigation of the electrical performance over the whole year will be the subject of a future paper.

Author Contributions: ManelHammami was the corresponding author and, with Simone Torretti, contributed to the manuscript composition, model developments, and thermal measurements. Simone Torretti also implemented and simulated the system by Simulink. Francesco Grimaccia and Gabriele Grandi supervised the manuscript composition and the experimental tests, carried out at the Department of Electrical, Electronic and Information Engineering, University of Bologna (Italy).

Conflicts of Interest: The authors declare no conflict of interest.

References

1. Thang, T.V.; Ahmed, A.; Kim, C.; Park, J. Flexible System Architecture of Stand-Alone PV Power Generation with Energy Storage Device. *IEEE Trans. Energy Convers.* **2015**, *30*, 1386–1396. [CrossRef]
2. Li, P.; Dargaville, R.; Cao, Y.; Li, D.-Y. Storage Aided System Property Enhancing and Hybrid Robust Smoothing for Large-Scale PV Systems. *IEEE Trans. Smart Grid* **2016**. [CrossRef]
3. Reddy, S.S.; Momoh, J.A. Realistic and Transparent Optimum Scheduling Strategy for Hybrid Power System. *IEEE Trans. Smart Grid* **2015**, *6*, 3114–3125. [CrossRef]
4. Fath, H.; Al Tarabsheh, A.; Ghazal, A.; Asad, M.; Morci, Y.; Etier, I.; El Haj, A. Performance of Photovoltaic Cells in Photovoltaic Thermal (PVT) Modules. *IET Renew. Power Gener.* **2016**, *10*, 1–7.
5. Khelifa, A.; Touafek, K.; Ben Moussa, H. Approach for the Modelling of Hybrid Photovoltaic–thermal Solar Collector. *IET Renew. Power Gener.* **2015**, *9*, 207–217. [CrossRef]
6. Xu, X.; Meyers, M.M.; Sammakia, B.G.; Murray, B.T.; Chen, C. Performance and Reliability Analysis of Hybrid Concentrating Photovoltaic/thermal Collectors with Tree-Shaped Channel Nets' Cooling System. *IEEE Trans. Compon. Packag. Manuf. Technol.* **2013**, *3*, 967–977. [CrossRef]
7. Grandi, G.; Rossi, C.; Hammami, M. *Modulo Fotovoltaico con Sistema di Accumulo Integrato (PV Module with Integrated Energy Storage)*; Patent Deposit Number A28517 LCA.gf; Alma Mater Studiorum, University of Bologna: Bologna, Italy, 2016.
8. Hu, Y.; Cao, W.; Ma, J.; Finney, S.J.; Li, D. Identifying PV Module Mismatch Faults by a Distribution Analysis. *IEEE Trans. Device Mater. Reliab.* **2014**, *14*, 951–960. [CrossRef]
9. Bardhi, M.; Grandi, G.; Tina, G.M. Comparison of PV Cell Temperature Estimation by Different Thermal Power Exchange Calculation Methods. In Proceedings of the International Conference on Renewable Energies and Power Quality (ICREPQ'12), Santiago de Compostela, Spain, 28–30 March 2012.
10. Tina, G.M.; Scrofani, S. Electrical and Thermal Model for PV Module Temperature Evaluation. In Proceedings of the Mediterranean Electrotechnical Conference, MELECON, Ajaccio, France, 5–7 May 2008; pp. 585–590.
11. Bardhi, M.; Grandi, G.; Premuda, M. Steady State Global Power Balance for Ground- Mounted Photovoltaic Modules Ground-Mounted Photovoltaic Modules. In Proceedings of the Third International Renewable Energy Congress, Hammamet, Tunisia, 20–22 December 2011; pp. 359–365.
12. Hemenway, D.; Sakurai, H.; Sampath, W.; Barth, K. Thermal Modeling of PV Modules Using Computational Simulation. In Proceedings of the IEEE 40th Photovoltaic Specialist Conference, Denver, CO, USA, 8–13 June 2014; pp. 1344–1347.
13. Chen, Y.; Zhuo, F.; Liu, X.; Xiong, L. Thermal Modelling and Performance Assessment of PV Modules Based on Climatic Parameters. In Proceedings of the IEEE Energy Conversion Congress and Exposition (ECCE), Montreal, QC, Canada, 20–24 September 2015; pp. 3282–3286.
14. Guyer, E.C. *Handbook of Applied Thermal Design*; CRC Press: Boca Raton, FL, USA, 1999.
15. Adrian, B. *Convective Heat Transfer*, 4th ed.; Wiley: Hoboken, NJ, USA, 2013.

16. Hu, Y.; Cao, W.; Wu, J.; Ji, B.; Holliday, D. Thermography-Based Virtual MPPT Scheme for Improving PV Energy Efficiency under Partial Shading Conditions. *IEEE Trans. Power Electron.* **2014**, *29*, 5667–5672. [CrossRef]
17. Romary, F.; Caldeira, A.; Jacques, S.; Schellmanns, A. Thermal Modelling to Analyze the Effect of Cell Temperature on PV Modules Energy Efficiency. In Proceedings of the 2011 14th European Conference on Power Electronics and Applications (EPE 2011), Birmingham, UK, 30 August–1 September 2011; pp. 1–9.

© 2017 by the authors. Licensee MDPI, Basel, Switzerland. This article is an open access article distributed under the terms and conditions of the Creative Commons Attribution (CC BY) license (http://creativecommons.org/licenses/by/4.0/).

Article

A Prototype Design and Development of the Smart Photovoltaic System Blind Considering the Photovoltaic Panel, Tracking System, and Monitoring System

Kwangbok Jeong [1], Taehoon Hong [1,*], Choongwan Koo [2], Jeongyoon Oh [1], Minhyun Lee [1] and Jimin Kim [3,4]

1 Department of Architecture & Architectural Engineering, Yonsei University, Seoul 03722, Korea; kbjeong7@yonsei.ac.kr (K.J.); omk1500@yonsei.ac.kr (J.O.); mignon@yonsei.ac.kr (M.L.)
2 Department of Building Services Engineering, Hong Kong Polytechnic University, Kowloon, Hong Kong, China; choongwan.koo@polyu.edu.hk
3 Division of Construction Engineering and Management, Purdue University, West Lafayette, IN 47907, USA; kim2752@purdue.edu
4 Department of Architecture & Architectural Engineering, Yonsei University, Seoul 03722, Korea
* Correspondence: hong7@yonsei.ac.kr; Tel.: +82-2-2123-5788

Received: 31 July 2017; Accepted: 13 October 2017; Published: 18 October 2017

Abstract: This study aims to design and develop the prototype models of the smart photovoltaic system blind (SPSB). To achieve this objective, the study defined the properties in three ways: (i) the photovoltaic (PV) panel; (ii) the tracking system; and (iii) the monitoring system. First, the amorphous silicon PV panel was determined as a PV panel, and the width and length of the PV panel were determined to be 50 mm and 250 mm, respectively. Second, the four tracker types (i.e., fixed type, vertical single-axis tracker, horizontal single-axis tracker, and azimuth-altitude dual-axis tracker) was applied, as well as the direct tracking method based on the amount of electricity generated as a tracking system. Third, the electricity generation and environmental conditions were chosen as factors to be monitored in order to evaluate and manage the technical performance of SPSB as a monitoring system. The prototype model of the SPSB is designed and developed for providing the electricity generated from its PV panel, as well as for reducing the indoor cooling demands through the blind's function, itself (i.e., blocking out sunlight).

Keywords: smart photovoltaic system blind; prototype model; photovoltaic panel; tracking system; monitoring system

1. Introduction

To solve the global warming potential and depletion of fossil fuels, the Paris agreement was adopted through the 21st Conference of the Parties held in Paris, France in December 2015. In response to this, the South Korea government established the '*Nationally Determined Contributions*' to reduce GHG emissions by 37% below the business-as-usual emission level by 2030 [1–4]. Accordingly, energy reduction is required in the building sector, which accounts for about 40% of the total fossil fuel consumption [5,6]. The South Korea government established the policy '*Obligation to Zero Energy Building by 2025*'. In a related move, the '*4th Renewable Energy Penetration Plan*' was established, and among the new and renewable energy sources, the ratio of photovoltaic (PV) system was increased from 2.7% (2015) to 4.1% (2030) [7–9]. However, there are certain limitations in looking to achieve the aforementioned goals using the rooftop PV system [10]. Thus, several previous studies were conducted to find the ways to install the PV system on the building façade, as well as rooftop. Menoufi et al. [11]

conducted the life cycle assessment of a building integrated concentrated PV scheme. Compared with the environmental impact of building integrated PV, that of a building integrated concentrated PV scheme, was analyzed to be lower. Tak et al. [12] designed a changeable organic semi-transparent solar cell window and analyzed its effect in terms of building energy efficiency and user comfort. In this context, the *'Act on the promotion of green buildings'* became effective in May 2015, which makes it compulsory to install shading devices in public buildings (office buildings and education facilities) whose total floor area exceeds 3000 m^2 with exterior walls that have windows or are made of glass [13]. The PV system blind can be considered one of the best solutions to provide the electricity generated from its PV panel and to reduce the indoor cooling demands through the blind's main function (i.e., blocking out sunlight) [14,15].

As shown in Table 1, previous studies concentrated on the design of the PV system and experiments on the building façades (vertical) as well as the building roofs (horizontal). In particular, these previous studies considered the properties of the PV system with regard three aspects (i.e., PV panel, tracking system, and monitoring system). The literature review can be summarized in three ways, as follows.

First, in terms of the PV panel, some of the previous studies have analyzed the technical performance of the PV system. Koo et al. [10], Mandalaki et al. [15], and Mandalaki et al. [16] analyzed the performance of the electricity generation with respect to the rooftop PV system and the shading devices with the integrated PV using the crystalline PV panel, respectively. Bahr [17] analyzed the PV blinds' optimal design parameters in terms of type of PV panel (i.e., crystalline silicon (c-Si) and a-Si) and installation options (i.e., ratio between the blinds installation distance to the module depth and tilt angle). Hwang et al. [18] conducted the optimization of the building integrated PV system considering four installation factors (i.e., module type, inclination, installation distance to module length ratio, and direction).

Second, in terms of the tracking system (i.e., tracker type and tracking method), some of the previous studies have analyzed the technical performance of the PV system. Lazaroiu et al. [19], Cruz-Peragón et al. [20], and Virtuani and Fanni [21], Heslop and Macgill [22], Mousazadeh et al. [23], and Dolara and Mussetta [24] performed comparative analyses on the electricity generation of the PV system according to the tracker type. In these previous studies, the performance of the electricity generation on the PV panel was found to be carried out in the order of the dual tracker, single tracker, and fixed type. Abdallah and Nijmeh [25] and Abdallah and Badran [26] developed a tracking system based on the direct tracking method and indirect tracking method, respectively. Through this, the performance of electricity generation due to the application of the tracking system was evaluated.

Third, in terms of the monitoring system, some of the previous studies have analyzed the technical performance of the PV system. Kang et al. [27] analyzed the relationship between the two variables by monitoring the electricity generation and temperature on the blinds integrated PV panels. Zahran et al. [28] developed a monitoring system for electricity generation and the temperature of PV surface by using Labview software developed by National Instruments (Austin, TX, USA). Kim et al. [29] and Kim et al. [30] performed the monitoring of the electricity generation and illuminance to analyze the electricity generation and incoming daylight from the PV blind.

Table 1. Literature review on the PV system.

Authors	Type of Application		PV panel			Tracking System		Properties of the PV System	Monitoring System			Type of Research		Country
	Rooftop (Horizontal)	Façade (Vertical)	c-Si	a-Si	CIGS	Tracker Type	Tracking Method	Electricity	Environmental Condition		Illuminance	Design	Experiment	
									Temperature	Solar Radiation				
Mandalaki et al. [16]	-	●	●	-	-	Fixed type	-	●	-	-	-	●	-	Greece
Koo et al. [10]	●	-	●	-	-	Fixed type	-	●	-	-	-	●	-	South Korea
Mandalaki et al. [15]	-	●	●	●	-	Fixed type	-	●	-	-	-	●	●	Greece
Bahr [17]	-	●	●	●	-	Fixed type	-	●	-	-	-	●	-	UAE
Hwang et al. [18]	●	●	●	●	-	Fixed type	-	●	-	-	-	●	-	South Korea
Lazaroiu et al. [19]	●	-	-	-	-	Fixed type, Single-axis tracker	-	●	-	-	-	-	●	-
Cruz-Peragón et al. [20]	●	-	-	●	-	Fixed type, Dual-axis tracker	-	●	-	-	-	●	-	Spain
Virtuani and Fanni [21]	●	-	●	●	-	Fixed type, Dual-axis tracker	-	●	-	-	-	-	●	Switzerland
Heslop and Macgill [22]	●	-	●	●	-	Fixed, Single-axis tracker, Dual-axis tracker	-	●	-	-	-	●	-	Australia
Mousazadeh et al. [23]	-	-	-	-	-	Fixed, Single-axis tracker, Dual-axis tracker	Direct tracking method	●	-	-	-	-	●	-
Dolara et al. [24]	●	-	●	-	-	Fixed type, Single-axis tracker	-	●	-	●	-	-	●	Italy
Abdallah and Nijmeh [25]	●	-	-	-	-	Fixed type, Dual-axis tracker	Direct tracking method	●	-	-	-	-	●	Jordan
Abdallah and Badran [26]	●	-	-	-	-	Fixed, Single-axis tracker	Indirect tracking method	●	●	●	-	-	●	Jordan
Kang et al. [27]	-	●	●	-	-	Single-axis tracker	-	●	●	-	-	●	-	South Korea
Zahran et al. [28]	●	-	-	-	-	-	-	●	●	-	-	-	●	-
Kim et al. [29]	-	●	●	-	-	Single-axis tracker	-	●	-	-	●	-	●	South Korea
Kim et al. [30]	-	●	●	-	-	Single-axis tracker	-	●	-	-	●	-	●	South Korea

Note: c-Si stands for the crystalline silicon PV panel; a-Si stands for the amorphous silicon PV panel; and CIGS stands for the copper-indium-gallium-selenide PV panel (CIGS: copper-indium-gallium-selenide).

As with the literature review, there are several limitations in the previous studies. First, the electricity generation of the c-Si PV panel and the a-Si PV panel were analyzed for the PV blind and rooftop PV system, whereas that of the copper-indium-gallium-selenide (CIGS) PV panel was not taken into account for the analysis. Second, as for the rooftop PV system, various tracker types (i.e., fixed, single-axis, and dual-axis tracker types) and tracking methods (i.e., direct and indirect tracking methods) were analyzed. On the other hand, as for the PV blind, only the fixed type and single-axis tracker type was analyzed and, accordingly, there is a lack of research that comprehensively considers the dual-axis tracker type and tracking method. Third, few studies are conducted to evaluate the feasibility of the PV blind that takes into consideration the three properties (i.e., PV panel, tracking system, and monitoring system) and performs the experimental research. To address these challenges, this study aimed to design and develop the prototype models of the smart photovoltaic system blind (SPSB) with consideration of the three properties (i.e., PV panel, tracking system, and monitoring system). The three properties (i.e., PV panel, tracking system, and monitoring system) of the SPSB were defined in Section 2. Additionally, detailed techniques applicable to the SPSB by property (i.e., PV panel: PV techniques and PV panel's design variables; tracking system: tracker type and tracking method; and monitoring system: monitoring factor) were explained. In Section 3, the developed three prototype models of the SPSB were described in terms of three properties (i.e., PV panel, tracking system, and monitoring system). Finally, the conclusion of this study was presented in Section 4.

2. A Prototype Design of the Smart Photovoltaic System Blind

To reduce the building energy, it is essential to develop a PV system that can be applied not only to the rooftop areas of buildings, but also to building façades. The proposed SPSB is a product that can be applied to building façades, especially to the external window areas of buildings. The major properties of the SPSB are composed of three characteristics: (i) the PV panel; (ii) the tracking system, and (iii) the monitoring system (refer to Figure 1). The PV panel generates the electricity through photoelectric effects using solar radiation. The tracking system helps maximize the electricity generation of the SPSB by controlling the angle of incidence between the PV panel and the incoming solar radiation using the printed circuit board and controller. The monitoring system establishes the current, voltage and power produced though the PV panel as a database and evaluates the optimal tilted and azimuth angle of the SPSB that maximizes the electricity generation. Based on the evaluation results, the tilted and azimuth angle of the PV panel is controlled to maintain the maximum electricity generation of the SPSB.

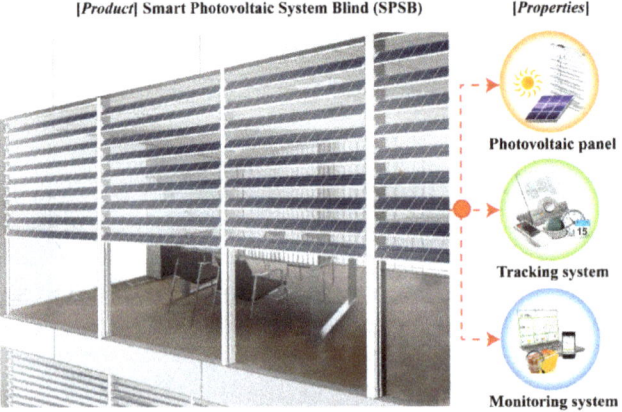

Figure 1. Concept of the smart photovoltaic system blind.

The SPSB is capable of making a variety of combinations of these three properties (i.e., PV panel, tracking system, and monitoring system) according to the particular circumstances and the user's decision. As a result, the technical performance of the SPSB (i.e., the amount of electricity generation) is determined. Therefore, the terminology "smart" used in this study signifies that the SPSB can generate maximum electricity output by combining the three properties [13].

2.1. Properties 1: PV panel

In terms of the PV panel properties, two factors were considered for the development of the SPSB in this study: (i) PV techniques applicable to the SPSB and (ii) PV panel designs applicable to the SPSB.

2.1.1. PV Techniques Applicable to the SPSB

The PV techniques applicable to the SPSB are largely divided into two generations: (i) the first-generation PV techniques (i.e., mono-crystalline silicon (mono-Si) and poly-crystalline silicon (poly-Si)); and (ii) the second-generation PV techniques (a-Si, CIGS, and cadmium telluride (CdTe) thin-film). The third-generation PV techniques (i.e., dye-sensitized PV panel (DSSP), organic PV (OPV) panel, and nano PV panel) were excluded from this study since the technology is still in an experimental stage [13,31–36]. To determine the PV techniques applicable to the SPSB, this study considered two main issues: (i) usability issues (i.e., efficiency, shading effect, and certification) and (ii) constructability issues (i.e., width, length, thickness, and weight) (refer to Table 2) [37–41].

Table 2. Usability and constructability issues of the first- and second-generation PV panels.

Classification			First-Generation PV Panels		Second-Generation PV Panels		
			Mono-Si	Poly-Si	a-Si	CIGS	CdTe
Usability issue	Efficiency (module level)		14.5%	14.0%	5.8%	10.5%	9.9%
	Shading effect		High	High	Low	Low	Low
	Harmful effect		No	No	No	No	Yes
Constructability issue	Size	Width	Customized	Customized	Customized	Customized	Customized
		Length	Customized	Customized	Customized	Customized	Customized
		Thickness	35 mm	48 mm	2.6 mm	2.5 mm	27.9 mm
	Weight		11.7 kg/m^2	11.6 kg/m^2	2.5 kg/m^2	2.4 kg/m^2	16.7 kg/m^2

Note: Mono-Si stands for the mono-crystalline silicon PV panel; Poly-Si stands for the poly-crystalline silicon PV panel; a-Si stands for the amorphous silicon PV panel; CIGS stands for the copper-indium-gallium-selenide PV panel; and CdTe stands for the cadmium telluride PV panel.

(i) Usability issues of the PV panel: The usability issues of the SPSB took into account three factors (i.e., shading effect, efficiency, and certification). First, the efficiency of the PV panel is a factor determined by the photoelectric effect—the conversion of solar radiation into electricity. Additionally, as the photoelectric effect of PV cell is greater, the efficiency of the PV panel increases. As shown in Table 2, the efficiency of the PV panel was determined to be the average efficiency of PV panel in module level (i.e., mono-Si: 14.5%, poly-Si: 14.0%, a-Si: 5.8%, CIGS: 10.5%, and CdTe: 9.9%) [42]. The efficiency of the second-generation PV panel (a-Si: 5.8%, CIGS: 10.5%, and CdTe: 9.9%) was determined to be 1.33–2.50 times inferior to that of the first-generation PV panel (i.e., mono-si: 14.5% and poly-si: 14.0%). However, unlike the first-generation PV panel, the second-generation PV panel exhibits excellent applicability due to the application of the thin film PV technique [35,42]. Second, the shading effect of the PV panel is a factor that has a significant influence on the electricity generation performance of the PV panel, and there is a noticeable difference in the sensitivity depending on the type of PV panel. As shown in Table 2, the second-generation PV panels (a-Si, CIGS, and CdTe: low) are superior to the first-generation PV panels (mono-Si and poly-Si: high) in terms of the shading effect [17,43,44]. Thus, with regard the shading effect of the PV panel, the second-generation PV panels were determined as a PV technique applicable to the SPSB. Third, the possible harmful effects of the PV panel were a factor determined depending on whether or not the toxic material may be used. According to the *Enforcement Rule of the 'Act on the promotion of the development, use and diffusion of new and renewable energy'* in South Korea, CdTd PV panels which are manufactured using toxic materials (i.e., cadmium and tellurium) were excluded from these considerations [45,46].

(ii) Constructability issues of the PV panel: The constructability issues of the SPSB took into account two factors (i.e., size and weight). Towards this end, this study considered the PV technique applicable to the SPSB based on the size and weight of the wood venetian blind as it is superior when compared to other types of blinds in terms of both size and weight [47]. First, in terms of size, the first and second-generation PV panels met the standards (i.e., width: below 63.5 mm, length: below 2438 mm, and thickness: below 3 mm) of the venetian blind. For example, in terms of width, the maximum width at which the wood venetian blind can be manufactured is 63.5 mm, and both the first and second-generation PV techniques (mono-Si, poly-Si, a-Si, and CIGS: customized) satisfy this criteria. Second, in terms of weight, the first-generation PV panel (mono-Si: 11.7 kg/m^2 and poly-Si: 11.6 kg/m^2) does not meet the standard (i.e., weight: 3 kg/m^2) of the venetian blind, whereas the second-generation PV panel (a-Si: 2.5 kg/m^2 and CIGS: 2.4 kg/m^2) does. Thus, in terms of the constructability issues of the PV panel, the second-generation PV panels were chosen as the PV technique applicable to the SPSB [48].

2.1.2. PV Panel's Design Variables Applicable to the SPSB

To determine the PV panel design applicable to the SPSB, the optimal size capable of maximizing the electricity generation of PV panels should be considered. Towards this end, three factors were considered in this study: (i) the length of the PV panel; (ii) the width of the PV panel; and (iii) the distance between the centerline of the SPSB and the exterior window. This is because the influence of the shading effect caused by the slat of the SPSB should be minimized to maximize the electricity generation of the SPSB.

(i) Length of the PV panel: According to previous studies, the length of the PV panel has little effect on the electricity generation per unit area. However, the length of the PV should be considered as a factor affecting the range of tracking [27].

(ii) Width of the PV panel: According to previous studies, the width of the PV panel is a factor that has a great effect on the electricity generation per unit area. This is because as the width of the PV panel increases, the shading effect caused by the upper slat increases. Thus, there is a need to determine the optimal width of the PV panel in order to maximize electricity generation [27].

(iii) Distance between the centerline of the SPSB and the exterior window: According to previous studies, the distance between the centerline of the SPSB and the exterior window is a factor that has an effect on the electricity generation per unit area [47]. This is because a farther distance between the SPSB and exterior window leads to reduced solar radiation, resulting in the reduction of electricity generation [27].

2.2. Properties 2: Tracking System

In terms of the tracking system properties, the two factors were considered for the development of the SPSB in this study: (i) the tracker type and (ii) the tracking method. First, the tracker type can take account of the fixed type, single-axis, and dual-axis tracker. Second, the tracking method can take account of the direct tracking method and the indirect tracking method [23,49].

2.2.1. Tracker Types Applicable to the SPSB

The tracker types should be considered to maximize electricity generation by increasing the solar radiation that reaches the PV panel. These tracker types can be categorized into the single-axis tracker and dual-axis tracker (refer to Figure 2).

(i) Single-axis tracker type: The single-axis tracker type is a technology that has one axis of rotation and can be divided into four tracker types: (1) the horizontal single-axis tracker (HSAT); the vertical single-axis tracker (VSAT); (2) the tilted single-axis tracker (TSAT); and (3) the polar aligned single-axis tracker (PASAT). The HSAT can track the daily north-south motion of the sun through the axis of rotation horizontal to the ground (refer to tracker type (A) in Figure 2). The VSAT can track the daily east-west motion of the sun through the axis of rotation vertical to the ground (Refer to tracker type (B) in Figure 2). The TSAT sets the axes of rotation between the horizontal and the vertical. Thus, the daily east-west motion of the sun is tracked by setting the axis of rotation to be parallel to the axis of the earth's rotation (refer to tracker type (C) in Figure 2). The PASAT is aligned to the polar star. Through this, the tilt angle becomes equal to the latitude of installation and allows it to align with the Earth's axis (refer to tracker type (D) in Figure 2) [23,49,50].

(ii) Dual-axis tracker type: The dual-axis tracker type is a technology that has two axes of rotation and can be divided into two tracker types: (1) the tip-tilt dual-axis tracker (TTDAT); and (2) the azimuth-altitude dual-axis tracker (AADAT). In general, the TTDAT tracks the daily east-west motion of the sun and north-south motion of the sun by rotating the PV panel fixed to the top of the pole (refer to tracker type (E) in Figure 2). The AADAT can track the daily east-west motion of the sun and the daily north-south motion of the sun using a large ring mounted on the ground with a series of rollers (refer to tracker type (F) in Figure 2) [23,49,50].

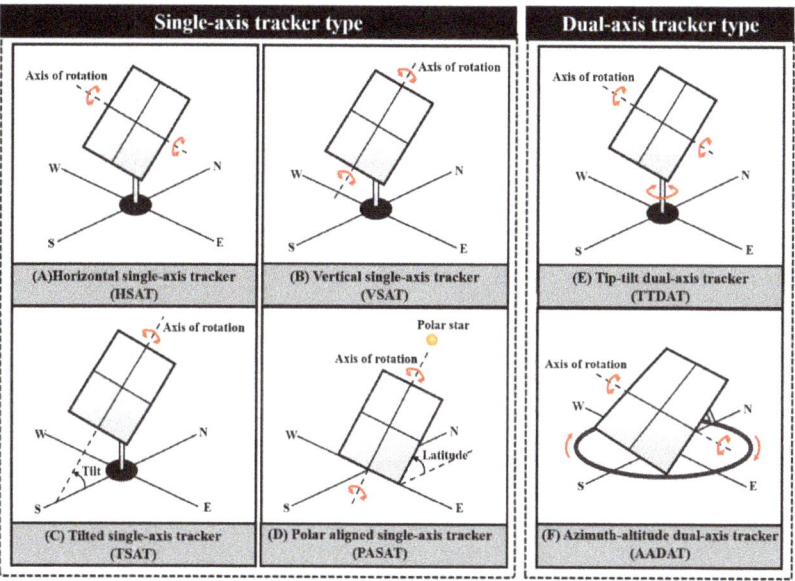

Figure 2. Single- and dual-axis tracker types.

2.2.2. Tracking Methods Applicable to the SPSB

According to control strategies, the tracking methods can be categorized into two types: (i) the direct tracking method; and (ii) the indirect tracking method.

(i) Direct tracking method: The direct tracking method is a method for controlling the tracking system based on the results of the direct measurement from the PV system, and the measurement data can be divided into two types (i.e., solar radiation and electricity generation). The solar radiation-based direct tracking method determines the tracking direction by using a photo-sensor and evaluating the maximum solar radiation data. The electricity generation-based direct tracking method determines the tracking direction based on the evaluation of the current data produced from the PV system. However, the direct tracking method has a disadvantage in that malfunction in tracking can occur as a result of temporary errors [51,52].

(ii) Indirect tracking method: The indirect tracking method is for controlling the tracking system by calculating the position of the sun according to both the date and time. However, this method has disadvantages in that it is difficult to reflect the difference of the electricity generation depending on the location and direction where the PV system is installed.

2.3. Properties 3: Monitoring System

The purpose of the monitoring system is to evaluate the technical performance of the SPSB (i.e., the amount of electricity generation) and to establish the maintenance strategy on the SPSB. Toward this end, the following two factors should be monitored and established as a database: (i) electricity generation; and (ii) environmental conditions.

(i) Electricity generation: In order to evaluate the technical performance of the SPSB, information regarding the amount of electricity generated (i.e., current, voltage, and power) is monitored and established as a database. With the use of this monitoring information, especially on the electricity generation, the electricity generation-based direct tracking method can be applied [27–30,51,52].

(ii) Environmental conditions: The SPSB is a technology based on a variety of electronic components, and it is affected by various external elements. In particular, the environmental conditions (i.e., temperature, humidity, and solar radiation) are a factor that has a very large influence on the performance of the SPSB. Accordingly, information regarding the environmental conditions (i.e., temperature, humidity, and solar radiation) should be monitored and then established as a database in order to perform effective maintenance management on the SPSB [26–28].

3. Development of Prototype Models of the Smart Photovoltaic System Blind

As shown in Figure 3, this study has developed three prototype models of the SPSB to improve the technical performance of the SPSB. The first prototype model was developed to evaluate the possibility of technical realization of the three properties (i.e., PV panel, tracking system, and monitoring system) in a form of blind. The frame and control system of the first prototype model were developed by using the 'Lego Product (e.g., Lego brick, Lego Mindstorms, etc.)' and Labview software, respectively. The second prototype model was developed to achieve a technology level for a real product through integrating these three properties. The third prototype model was developed to improve the technology level of the SPSB for applying it to the real building façade. The frame and control system of the second and third prototype model were developed by using materials actually used to make the real window frames and Labview software, respectively.

Meanwhile, this study considered the three properties to develop the three prototype model. As shown in Table 3, different technologies were applied by properties according to the purpose of the prototype model. The detailed explanations of the prototype model with regard the three properties (i.e., PV panel, tracking system, and monitoring system) are presented in Section 3.

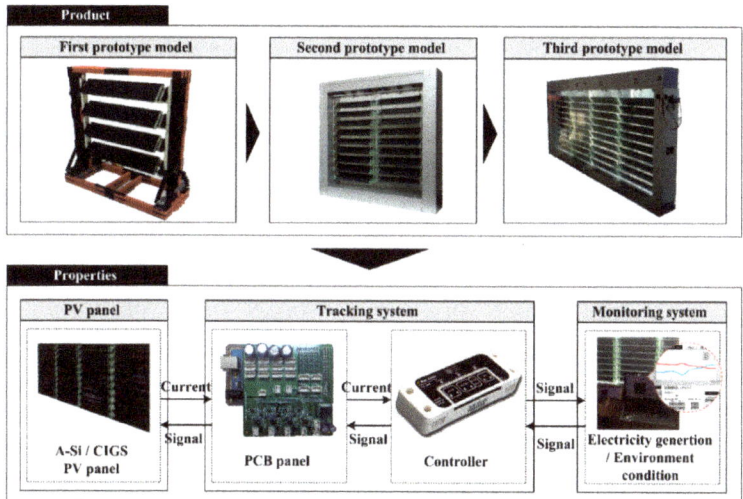

Figure 3. The prototype models of the SPSB and the associated properties.

Table 3. Characteristic by the prototype model of the SPSB.

Properties	Classification		First Prototype Model	Second Prototype Model	Third Prototype Model
PV panel	a-Si		-	•	•
	CIGS		•	-	-
Tracking system	Tracker type	Single	VSAT	HSAT, VSAT	HSAT, VSAT
		Dual	-	AADAT	AADAT
	Tracking method	Direct	•	•	•
		Indirect	-	-	-
Monitoring system	Electricity generation	Current	•	•	•
		Voltage	-	•	•
		Power	-	•	•
	Environmental condition	Temperature	-	-	•
		Humidity	-	-	•

Note. a-Si stands for the for the amorphous silicon PV panel; CIGS stands for the copper-indium-gallium-selenide PV panel; VSAT stands for the vertical single-axis tracker; HSAT stands for the horizontal single-axis tracker; and AADAT stands for the azimuth-altitude dual-axis tracker.

3.1. Properties 1: PV panel

3.1.1. PV Techniques Installed in Prototype Models of the SPSB

In consideration of the PV techniques applicable to the SPSB (i.e., usability and constructability issues), the a-Si and CIGS PV panels which are the second-generation PV panels were used to develop the prototype models of the SPSB (refer to Table 3). As shown in Table 2, the efficiency of the CIGS PV panel is superior to that of the a-Si PV panel. The CIGS PV panel was used in the first prototype model, whereas the a-Si PV panel was used in the second and third prototype models. The reason for changing from the CIGS PV panel to a-Si PV panel is as follows. The a-Si mini PV panel which can be applied to the SPSB prototype model can be produced as a product, while the CIGS mini PV panel cannot. This is because, unlike the a-Si PV panel, the CIGS PV panel, which applies a roll-to-roll manufacturing process, cannot be produced in the form of mini CIGS PV panel which can be applied to the SPSB prototype model. Thus, a-Si PV panel was used in the second and third prototype models to guarantee the stability and reliability of electricity generation measurement results and the supply of components for developing the prototype model of the SPSB. Specific details for each prototype model are as follows.

The first prototype model consisted of a total of four blind slats, and the 'SP1-75' CIGS PV panels manufactured (maximum power: 75 W) by Solopower were attached to each blind slat (refer to Table 4) [53,54]. The 'SP1-75' CIGS PV panels were selected as products that can be purchased in South Korea. Since the 'SP1-75' CIGS PV panel is too large to be applied to the first prototype model, 'SP1-75' CIGS PV panels were cut to be applicable size for the blind slat of the first prototype model. In addition, CIGS PV panel attached to all blind slats were linked in parallel. The second and third prototype models of the SPSB used the '502500-16V' a-Si PV panel manufactured by Solar Center (refer to Table 4). Furthermore, a-Si PV panels attached to all of the blind slats were linked in parallel. The second prototype model of the SPSB was composed of nine blind slats per column and had a total of two columns and 18 blind slats. The third prototype model of the SPSB was composed of ten blind slats per column and had a total of two columns and 40 blind slats.

Table 4. Specifications of the PV panel by prototype model.

Classifications	First Prototype Model	Second Prototype Model	Third Prototype Model
PV panel	SP1-75	502500-16V	502500-16V
Manufacturer	Solopower	Solar Center	Solar Center
Size of PV panel	398 mm × 2197 mm × 2 mm	250 mm × 50 mm × 2 mm	250 mm × 50 mm × 2 mm
Maximum power (P_{max})	75 W	1.5 W	1.5 W
Maximum power voltage (V_{mp})	21.8 V	16 V	16 V
Optimum operating current (I_{mp})	3.4 A	95 mA	95 mA
Short-circuit current (I_{sc})	4.3 A	99 mA	99 mA
Open-circuit voltage (V_{oc})	30.6 V	19.4 V	19.4 V

3.1.2. PV Panel's Design Variables Installed in Prototype Models of the SPSB

As mentioned in Section 2.1.2, the three factors (i.e., length of the PV panel, width of the PV panel, and distance between the centerline of the SPSB and the exterior window) were considered for the development of the prototype models of the SPSB.

The length of the SPSB can be adjusted according to the circumstance and user's decision. However, the length of the prototype models was determined based on the size of the PV panel to guarantee the stability and reliability of the electricity generation measurement results and the supply of components for developing the prototype model of the SPSB. Accordingly, as shown in Table 4, the lengths of the PV panel of the first, second, and third prototype models were determined to be 398 mm, 250 mm, and 250 mm, respectively, based on the length of the 'SP1-75' CIGS PV panel and the '502500-16V' a-Si PV panel [53,54]. Second, the width of the PV panel applied to the SPSB can be determined according to the blind type (e.g., venetian blind, pleated blind, etc.). In this study, the width of the PV panel was determined to be 50 mm based on the size of a standard venetian blind [47]. As mentioned in Section 2.1.2, the greater distance between the SPSB and exterior window leads to the degradation of the electricity generation performance of the SPSB. In this regard, the distance between the centerline of the SPSB and the exterior window was determined to be 25 mm [27].

3.2. Properties 2: Tracking System

3.2.1. Tracker Types Installed in Prototype Models of the SPSB

According to previous studies, the performance of the electricity generation of the PV system was enhanced in the order of the dual-axis tracker type, single-axis tracker type, and fixed type [19–22]. To improve the performance of the electricity generation of the SPSB, the application of the tracker is required. Therefore, among the total five tracker types, four tracker types (i.e., fixed type, VSAT, HSAT, and AADAT) that can track the daily east-west motion and north-south motion of the sun were applied to the prototype model of the SPSB.

The first prototype model applied the VSAT to track the north-south motion of the sun. In the second and third prototype models, the four tracker types (i.e., fixed type, VSAT, HSAT, and AADAT) that can track the daily east-west motion and north-south motion of the sun were all applied. Figure 4 shows the concept diagram of the third prototype model of the SPSB. All the PV panels on the blind slat are connected each other with lifting and rotating ropes. The prototype model of the SPSB moves along the north-south motion of the sun using the motor 'A', motor 'B', rotating rope, and wire-puller, whereas the prototype model of the SPSB moves along the east-west motion of the sun using motor 'C' and gears. Aforementioned in Section 2.2.1, the VSAT and HSAT can track the daily east-west motion (X-axis) and north-south motion (Y-axis) of the sun, respectively. In addition, the AADAT can track the daily east-west motion (X-axis) and north-south motion (Y-axis) of the sun. In particular, since the third prototype model is composed of four columns, a comparative analysis on the performance of the electricity generation of the SPSB can be conducted by applying the four tracker types (i.e., fixed type, VSAT, HSAT, and AADAT) at the same time.

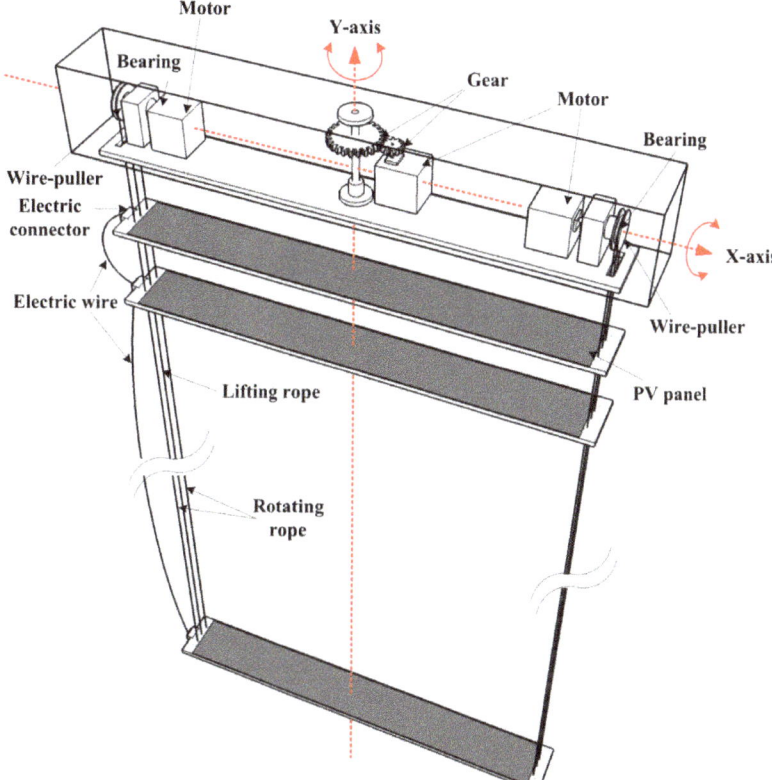

- VSAT (vertical single-axis (X-axis) tracker): Track the daily east-west motion of the sun.
- HSAT (horizontal single-axis (Y-axis) tracker): Track the daily north-south motion of the sun.
- AADAT (azimuth-altitude dual-axis (X- and Y-axes) tracker): Track the daily east-west motion and daily north-south motion of the sun.

Figure 4. Concept diagram of the third prototype model of the SPSB.

3.2.2. Tracking Methods Installed in Prototype Models of the SPSB

Among the two tracking methods, the direct tracking method was used to develop the prototype model. Particularly with consideration of the costs, the direct tracking method based on the produced electricity generation was applied to the three prototype models of the SPSB.

Figure 5 shows the schematic circuit for measuring the electricity generation of the SPSB. Two factors were considered to measure the current of the prototype model: (i) amplification; and (ii) the low-pass filter.

First, the three prototype models produce small levels of electricity as a miniature representation of the SPSB. For example, the third-generation prototype model of the SPSB can produce a current of up to 950 mA (e.g., third prototype model: 95 mA × 10 module) per column. However, due to the performance of the PV panel and weather conditions (e.g., solar radiation, temperature, sky coverage, etc.), the electricity generation of this prototype model decreases. Accordingly, the amplification should be performed to measure the low current of the prototype model of the SPSB.

Second, the filtering on the amplified low current of the SPSB should be conducted using the low pass filter. The low-pass filter removes the noise amplified in the process of the amplification of the low current of the prototype model by blocking frequencies higher than the cutoff frequency and passing frequencies lower than the cutoff frequency [51].

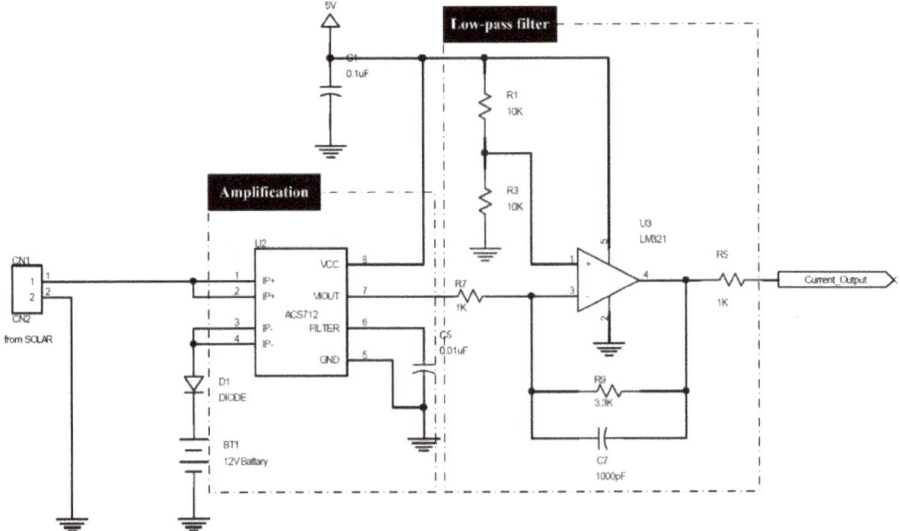

Figure 5. Schematic circuit for measuring the amount of electricity generated from the SPSB.

3.3. Properties 3 (Monitoring System)

The monitoring system of the first and second prototype model of the SPSB displays the time-series data only on the electricity generation (i.e., current, voltage, and power). However, the monitoring system of the third prototype model of the SPSB not only can establish a time-series database on the electricity generation but also the environmental conditions (i.e., temperature and humidity) and optimal tiled and azimuth angles (refer to Figure 6). Through this, the electricity generation depending on the tracker type and environmental conditions can be analyzed in areas where the SPSB is installed. As a result, the optimal tiled and azimuth angle according to the environmental condition can be determined.

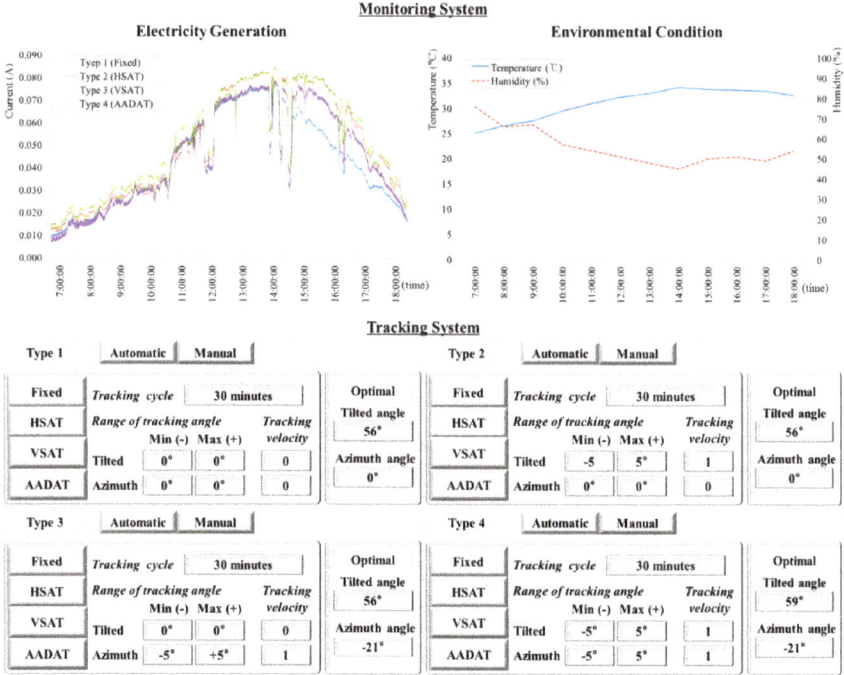

Figure 6. Graphical user interface of the monitoring system in the third prototype model.

3.4. Discussion

This study developed three prototype models of the SPSB to be installed on the building façade. It sought to improve the technical performance of the SPSB (i.e., the amount of electricity generation) in the development process of the three prototype models, starting from the first prototype model. In the most recently developed prototype model of the SPSB, the third prototype model, the following technologies were applied according to the three properties (i.e., PV panel, tracking system, and monitoring system): (i) a-Si PV panel; (ii) four tracker type (i.e., fixed type, VSAT, HSAT, and AADAT) and a tracking method (i.e., direct tracking method); and (iii) monitoring system (i.e., electricity generation and environment conditions).

The experimental studies were conducted using the third prototype model of the SPSB. Figure 7 shows the electricity generation of the third prototype model by tracker type (i.e., fixed type, VSAT, HSAT, and AADAT) (6 August 2015). First, the electricity generation of the third prototype model shows the inverted U-shape (with the highest value at 14:00 (0.80~1.02 W) and a lower value at 7:00 (0.01–0.03 W) and at 18:00 (0.04–0.09 W)). Additionally, the temperature and humidity show the inverted U-shape (highest value at 14:00 (34.2 °C) and the lowest value at 7:00 (25.1 °C)) and the U-shape (lowest value at 14:00 (45%) and the highest value at 7:00 (76%)), respectively. This is because the solar radiation in South Korea is highest at 14:00 (2.89 MJ/m^2) [55]. Second, the dual-axis tracker (i.e., AADAT) can improve the electricity generation of the third prototype model than other methods (single-axis tracker (i.e., HSAT and VSAT) and fixed type). Additionally, compared with the electricity generation of the third prototype model using VSAT, the electricity generation of the third prototype model using HSAT is improved. The VSAT applied to SPSB has a limitation for tracking the east-west motion of the sun. This is because SPSB installed in the window has a limitation on the depth.

Meanwhile, with the use of the third prototype model of the SPSB, the following experimental studies can be carried out. First, an evaluation on the electricity generation by tracking system (i.e., the tracker type and tracking method) according to the location of the SPSB installation can be performed. Second, in the maintenance phase, the optimal electricity generation strategy for maximum electricity output can be established using monitoring data such as electricity generation (i.e., current, voltage, and power) and environmental conditions (temperature, humidity, and solar radiation).

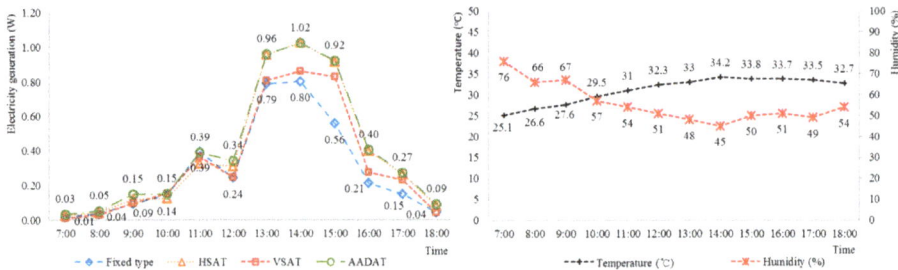

Figure 7. Electricity generation and the environmental condition of the third prototype model.

4. Conclusions

To cope with POST-2020, the South Korea government established the *'Nationally Determined Contributions'* to reduce GHG emissions by 37% below the business-as-usual emission level by 2030. The PV system blind can be considered one of the best solutions to provide electricity generated from its PV panel and to reduce the indoor cooling demands through the blind's main function. Thus, this study aimed to design and develop the prototype model of the SPSB. Toward this end, the study defined the properties in three ways (i.e., PV panel, tracking system, and monitoring system). It also developed three prototype models in order to improve the technical performance of the SPSB. In the third prototype model, the following technologies were applied by properties. First, in terms of the PV techniques applicable to the SPSB, the a-Si PV panel was chosen. Second, in terms of the tracking system, the four tracker types (i.e., fixed type, VSAT, HSAT, and AADAT) and the direct tracking method based on electricity generation were applied. Third, in terms of the monitoring system, the electricity generation and environmental conditions were monitored to evaluate and manage the technical performance of the SPSB. The result of experimental study was conducted using the third prototype model of the SPSB. First, the electricity generation of the third prototype model shows the inverted U-shaped. Second, the electricity generation of the third prototype model is superior in order of the AADAT, HSAT, VSAT, and the fixed type applied to SPSB.

The prototype model of the SPSB is designed and developed for providing electricity generated from its PV panel, as well as for reducing the indoor cooling demands through the blind's main function (i.e., blocking out sunlight). Furthermore, it is expected that it will establish an optimal operation strategy to maximize the amount of electricity generated from the SPSB by considering the three properties. To improve the SPSB, future research can be employed to the computational intelligence techniques (e.g., artificial neural network, harmony search meta-heuristic algorithm, genetic algorithm, etc.) as follows: First, the electricity generation from the SPSB considering environmental conditions (e.g., orientation, number of story, weather, etc.) of the building will be forecasted for optimizing the optimal size of the SPSB applicable to the building. For that, the use of the time-series historical data of the electricity and environment conditions makes it possible to forecast the electricity generation from the SPSB [56–58]. Second, the efficiency of the electricity generation can be improved through the hybrid tracking method. For that, the hybrid tracking method can be developed through a combination of various direct and indirect tracking methods using the computational intelligence techniques [59–61]. Third, in terms of the building energy supply and demand, the monitoring system of the SPSB will be developed using the computational intelligence techniques. The optimal operation strategy of the SPSB can be determined considering the electricity generation from the SPSB and electricity consumption from the electronics [62].

Acknowledgments: This research was supported by a grant (17CTAP-C117226-02) from the Technology Advancement Research Program (TARP) funded by the Ministry of Land, Infrastructure, and Transport of the South Korean government.

Author Contributions: Kwangbok Jeong, Taehoon Hong, Choongwan Koo, Jeongyoon Oh, Minhyun Lee, and Jimin Kim conducted the design of the prototype model of smart photovoltaic system blind; and Kwangbok Jeong, Taehoon Hong, and Choongwan Koo wrote the paper.

Conflicts of Interest: The authors declare no conflict of interest.

Nomenclature

AADAT	Azimuth-altitude dual-axis tracker
a-Si	Amorphous silicon
CdTe	Cadmium tell
CIGS	Copper-indium-gallium-selenide
C-Si	Crystalline silicon
DSSP	Dye-sensitized photovoltaic panel
HSAT	Horizontal single-axis tracker
Mono-Si	Mono-crystalline silicon
OPV	Organic photovoltaic panel
PASAT	Polar aligned single-axis tracker
Poly-Si	Poly- crystalline silicon
PV	Photovoltaic
SPSB	Smart photovoltaic system blind
TTDAT	Tip-tilt dual-axis tracker
TSAT	Tilted single-axis tracker
VSAT	Vertical single-axis tracker

References

1. Jeong, K.; Hong, T.; Ban, C.; Koo, C. Life Cycle Economic and Environmental Assessment for Establishing the Optimal Implementation Strategy of Rooftop Photovoltaic System in Military Facility. *J. Clean. Prod.* **2015**, *104*, 315–327. [CrossRef]
2. International Energy Agency (IEA). *Urban BIPV in the New Residential Construction Industry*; IEA: Ottawa, ON, Canada, 2008.
3. Joint Research Centre (JRC). *PV Status Report 2011*; JRC: Ispra, Italy, 2011.
4. Hong, T.; Koo, C.; Kwak, T. Framework for the implementation of a new renewable energy system in an educational facility. *Appl. Energy* **2013**, *103*, 539–551. [CrossRef]
5. U.S. Energy Information Administration (EIA). *Annual Energy Review 2011*; EIA: Washington, DC, USA, 2012.
6. Koo, C.; Kim, H.; Hong, T. Framework for the analysis of the low-carbon scenario 2020 to achieve the national carbon emissions reduction target: Focused on educational facilities. *Energy Policy* **2014**, *73*, 356–367. [CrossRef]
7. Korea Energy Management Corporation (KEMC). *New & Renewable Energy White Paper 2012*; KEMC: Gyeonggi-do, Korea, 2012.
8. Lee, M.; Koo, C.; Hong, T.; Park, H.S. Framework for the mapping of the monthly average daily solar radiation using an advanced case-based reasoning and a geostatistical technique. *Environ. Sci. Technol.* **2014**, *48*, 4604–4612. [CrossRef] [PubMed]
9. Hong, T.; Koo, C.; Kim, H.; Park, H.S. Decision support model for establishing the optimal energy retrofit strategy for existing multi-family housing complexes. *Energy Policy* **2014**, *66*, 157–169. [CrossRef]
10. Koo, C.; Hong, T.; Park, H.S.; Yun, G. Framework for the analysis of the potential of the rooftop photovoltaic system to achieve net-zero energy solar buildings. *Prog. Photovolt. Res. Appl.* **2014**, *22*, 462–478. [CrossRef]
11. Menoufi, K.; Chemisana, D.; Rosell, J.I. Life cycle assessment of a building integrated concentrated photovoltaic scheme. *Appl. Energy* **2013**, *111*, 505–514. [CrossRef]

12. Tak, S.; Woo, S.; Park, J.; Park, S. Effect of the changeable organic semi-transparent solar cell window on building energy efficiency and user comfort. *Sustainability* **2017**, *9*, 950. [CrossRef]
13. Ministry of Land Infrastructure and Transport (MOLIT). *Green Building Development Support Act*; MOLIT: Sejong, Korea, 2015.
14. Koo, C.; Hong, T.; Jeong, K.; Ban, C.; Oh, J. Development of the smart photovoltaic system blind and its impact on the net-zero energy solar buildings using technical-economic-policy analyses. *Energy* **2017**, *124*, 382–396. [CrossRef]
15. Mandalaki, M.; Tsoutsos, T.; Papamanolis, N. Integrated PV in shading systems for Mediterranean countries: Balance between energy production and visual comfort. *Energy Build.* **2014**, *77*, 445–456. [CrossRef]
16. Mandalaki, M.; Zervas, K.; Tsoutsos, T.; Vazakas, A. Assessment of fixed shading devices with integrated PV for efficient energy use. *Sol. Energy* **2012**, *86*, 2561–2575. [CrossRef]
17. Bahr, W. A comprehensive assessment methodology of the building integrated photovoltaic blind system. *Energy Build.* **2014**, *82*, 703–708. [CrossRef]
18. Hwang, T.; Kang, S.; Kim, J.T. Optimization of the building integrated photovoltaic system in office buildings—Focus on the orientation, inclined angle and installed area. *Energy Build.* **2012**, *46*, 92–104. [CrossRef]
19. Lazaroiu, G.C.; Longo, M.; Roscia, M.; Pagano, M. Comparative analysis of fixed and sun tracking low power PV systems considering energy consumption. *Energy Convers. Manag.* **2015**, *92*, 143–148. [CrossRef]
20. Cruz-Peragón, F.; Casanova-Peláez, P.J.; Díaz, F.A.; López-García, R.; Palomar, J.M. An approach to evaluate the energy advantage of two axes solar tracking systems in Spain. *Appl. Energy* **2011**, *88*, 5131–5142. [CrossRef]
21. Virtuani, A.; Fanni, L. Seasonal power fluctuations of amorphous silicon thin-film solar modules: Distinguishing between different contributions. *Prog. Photovolt. Res. Appl.* **2014**, *22*, 208–217. [CrossRef]
22. Heslop, S.; MacGill, I. Comparative analysis of the variability of fixed and tracking photovoltaic systems. *Sol. Energy* **2014**, *107*, 351–364. [CrossRef]
23. Mousazadeh, J.; Keyhani, A.; Javadi, A.; Mobli, H.; Abrinia, K.; Sharifi, A. A review of principle and sun-tracking methods for maximizing solar systems output. *Renew. Sustain. Energy Rev.* **2009**, *13*, 1800–1818. [CrossRef]
24. Dolara, A.; Grimaccia, F.; Leva, S.; Mussetta, M.; Faranda, R.; Gualdoni, M. Performance analysis of a single-axis tracking PV system. *IEEE J. Photovolt.* **2012**, *2*, 524–531. [CrossRef]
25. Abdallah, S.; Nijmeh, S. Two axes sun tracking system with PLC control. *Energy Convers. Manag.* **2004**, *45*, 1931–1939. [CrossRef]
26. Abdallah, S.; Badran, O.O. Sun tracking system for productivity enhancement of solar still. *Desalination* **2008**, *220*, 669–676. [CrossRef]
27. Kang, S.; Hwang, T.; Kim, J.T. Theoretical analysis of the blinds integrated photovoltaic modules. *Energy Build.* **2012**, *46*, 86–91. [CrossRef]
28. Zahran, M.; Atia, Y.; Al-Hussain, A.; El-Sayed, I. LabVIEW Based Monitoring System Applied for PV Power Station. In Proceedings of the 12th WSEAS International Conference on Automatic Control, Modelling & Simulation, Catania, Italy, 29–31 May 2010.
29. Kim, S.H.; Kim, I.T.; Choi, A.S.; Sung, M.K. Evaluation of optimized PV power generation and electrical lighting energy savings from the PV blind-integrated daylight responsive dimming system using LED lighting. *Sol. Energy* **2014**, *107*, 746–757. [CrossRef]
30. Kim, J.J.; Jung, S.K.; Choi, Y.S.; Kim, J.T. Optimization of photovoltaic integrated shading devices. *Indoor Built Environ.* **2009**, *19*, 114–122. [CrossRef]
31. Park, H.; Koo, C.; Hong, T.; Oh, J.; Jeong, K. A finite element model for estimating the technical-economic performance of the building-integrated photovoltaic blind. *Appl. Energy* **2016**, *179*, 211–227. [CrossRef]
32. Korea Energy Management Corporation (KEMCO). *New & Renewable Energy White Paper*; Korea Energy Management Corporation (KEMCO): Yongin, Korea, 2014.
33. Korea Communications Agency (KCA). *Trends and Prospects of CIGS Thin-Film Photovoltaic Cell Technology*; Korea Communications Agency (KCA): Naju, Korea, 2014.
34. Korea Institute for Industrial Economics & Trade (KIET). *Competition Structure Analysis and Cooperation Plan of Photovoltaic Industry between Korea and China*; Korea Institute for Industrial Economics & Trade (KIET): Sejong, Korea, 2013.

35. Park, C. *Market Trend of Thin-Film Photovoltaic Cell by the Technology Type, and Expectation of Commercialization*; Market Report; Korea Institute of Science and Technology Information (KISTI): Daejeon, South Korea, 2014; Volume 4, pp. 3–6.
36. Energy Informative: Solar Panels and Home Energy Efficiency. Available online: http://energyinformative.org (accessed on 27 July 2015).
37. Hong, T.; Han, S.; Lee, S. Simulation-based determination of optimal life-cycle cost for FRP bridge deck panels. *Autom. Constr.* **2007**, *16*, 140–152. [CrossRef]
38. Han, S.; Hong, T.; Lee, S. Production prediction of conventional and global positioning system-based earthmoving systems using simulation and multiple regression analysis. *Can. J. Civ. Eng.* **2008**, *35*, 574–587. [CrossRef]
39. Hong, T.; Hastak, M. Simulation study on construction process of FRP bridge deck panels. *Autom. Constr.* **2007**, *16*, 620–631. [CrossRef]
40. Koo, C.W.; Hong, T.H.; Hyun, C.T.; Park, S.H.; Seo, J.O. A study on the development of a cost model based on the owner's decision making at the early stages of a construction project. *Int. J. Strateg. Prop. Manag.* **2010**, *14*, 121–137. [CrossRef]
41. Hong, T.; Jeong, K.; Koo, C.; Kim, J.; Lee, M. A preliminary study for determining photovoltaic panel for a smart photovoltic blind considering usability and constructability issues. *Energy Procedia* **2016**, *88*, 363–367. [CrossRef]
42. Energy Informative. Available online: http://energyinformative.org/OLD/wp-content/themes/nexus/sort.php?q=&manfilter=all&csifilter=all&cellfilter=Mono&min=&max=&man=1&mod=1&cell=1&eff=1&amountofitems=100 (accessed on 1 May 2015).
43. Khaing, H.H.; Liang, Y.J.; Htay, N.N.M.; Fan, J. Characteristics of Different Solar PV Modules under Partial Shading. *Int. J. Electr. Comput. Electr. Commun. Eng.* **2014**, *8*, 1328–1332.
44. Dolara, A.; Lazaroiu, G.C.; Leva, S.; Manzolini, G. Experimental investigation of partial shading scenarios on PV (photovoltaic) modules. *Energy* **2013**, *55*, 466–475. [CrossRef]
45. Ministry of Trade Industry & Energy (MOTIE). *Enforcement Rule of Act on the Promotion of the Development, Use and Diffusion of New and Renewable Energy*; MOTIE: Sejong, Korea, 2015.
46. Korea Energy Management Corporation (KEMCO). *Detaild Standard for the Examination of New and Renewable Energy Equipment*; KEMCO: Gyeonggi, Korea, 2014.
47. Levolor Window Fashions (LWF). *Wood Blind Specification 2005*; LWF: Rockaway, NJ, USA, 2005.
48. FreeCleanSolar. Available online: http://www.freecleansolar.com/default.asp (accessed on 15 September 2017).
49. Agee, J.T.; Obok-Opok, A.; Lazzer, M.D. Solar tracker technologies: Market trends and field applications. *Adv. Mater. Res.* **2007**, *18–19*, 339–344. [CrossRef]
50. Drury, E.; Lopez, A.; Denholm, P.; Margolis, R. Relative performance of tracking versus fixed tilt photovoltaic systems in the USA. *Prog. Photovolt. Res. Appl.* **2014**, *22*, 1302–1315. [CrossRef]
51. Salas, V.; Olías, E.A.; Barrado, A.; Lázaro, A. Review of the maximum power point tracking algorithms for stand-alone photovoltaic systems. *Sol. Energy Mater. Sol. Cells* **2006**, *90*, 1555–1578. [CrossRef]
52. Ma, J.; Man, K.L.; Ting, T.O.; Zhang, N.; Lei, C.U.; Wong, N. A hybrid MPPT method for photovoltaic system via estimation and revision method. In Proceedings of the 2013 IEEE International Symposium on Circuits and Systems, Beijing, China, 19–23 May 2013.
53. SoloPower Systems (SPS). SP1 Specifications. Available online: http://solopower.com/products/solopower-sp1 (accessed on 5 March 2016).
54. Solar Center (SC). a-Si Specifications. Available online: http://www.solarcenter.co.kr/dir/home/2009013085ryeeHVdc/home.php?go=shop_item_list&subj_code=003003000 (accessed on 5 March 2016).
55. National Climate Data Service System (NCDSS). Available online: http://sts.kma.go.kr/eng/jsp/home/contents/main/main.do (accessed on 15 September 2017).
56. Grimaccia, F.; Leva, S.; Mussetta, M.; Ogliari, E. ANN sizing procedure for the day-ahead output power forecast of a PV plant. *Appl. Sci.* **2017**, *7*, 622. [CrossRef]
57. Chow, S.; Lee, E.; Li, D. Short-term prediction of photovoltaic energy generation by intelligent approach. *Energy Build.* **2012**, *55*, 660–667. [CrossRef]
58. Ogliari, E.; Dolara, A.; Manzolini, G.; Leva, S. Physical and hybrid methods comparison for the day ahead PV output power forecast. *Renew. Energy* **2017**, *113*, 11–21. [CrossRef]

59. Guo, M.; Zang, H.; Gao, S.; Chen, T.; Xiao, J.; Cheng, L.; Wei, Z.; Sun, G. Optimal tilt angle and orientation of photovoltaic modules using HS algorithm in different climates of China. *Appl. Sci.* **2017**, *7*, 1028. [CrossRef]
60. Yousef, H. Design and implementation of a fuzzy logic computer-controlled sun tracking system. In Proceedings of the IEEE International Symposium on Industrial Electronics, Bled, Slovenia, 12–16 July 1999; Volume 3, pp. 1030–1034.
61. Luque, I.; Gordillo, F.; Rodriguez, F. PI based hybrid sun tracking algorithm for photovoltaic concentration. In Proceedings of the IEEE 19th European Photovoltaic Energy Conversion, Paris, France, 7–14 June 2004; pp. 7–14.
62. Hong, Y.; Yo, P. Novel genetic algorithm-based energy management in a factory power system considering uncertain photovoltaic energies. *Appl. Sci.* **2017**, *7*, 438. [CrossRef]

© 2017 by the authors. Licensee MDPI, Basel, Switzerland. This article is an open access article distributed under the terms and conditions of the Creative Commons Attribution (CC BY) license (http://creativecommons.org/licenses/by/4.0/).

Article

Optimal Tilt Angle and Orientation of Photovoltaic Modules Using HS Algorithm in Different Climates of China

Mian Guo [1], Haixiang Zang [1,2,*], Shengyu Gao [3], Tingji Chen [3], Jing Xiao [3], Lexiang Cheng [3], Zhinong Wei [1] and Guoqiang Sun [1]

[1] College of Energy and Electrical Engineering, Hohai University, Nanjing 211100, China; hhuguomian@163.com (M.G.); wzn_nj@263.net (Z.W.); hhusunguoqiang@163.com (G.S.)
[2] Jiangsu Key Laboratory of Smart Grid Technology and Equipment, Nanjing 210096, China
[3] State Grid Nanjing power supply company, Nanjing 210019, China; gaosy@js.sgcc.com.cn (S.G.); ctj@js.sgcc.com.cn (T.C.); xiaojing@js.sgcc.com.cn (J.X.); chenglx@js.sgcc.com.cn (L.C.)
* Correspondence: zanghaixiang@hhu.edu.cn; Tel.: +86-137-7071-9919

Received: 6 September 2017; Accepted: 29 September 2017; Published: 6 October 2017

Abstract: Solar energy technologies play an important role in shaping a sustainable energy future, and generating clean, renewable, and widely distributed energy sources. This paper determines the optimum tilt angle and optimum azimuth angle of photovoltaic (PV) panels, employing the harmony search (HS) meta-heuristic algorithm. In this study, the ergodic method is first conducted to obtain the optimum tilt angle and the optimum azimuth angle in several cities of China based on the model of Julian dating. Next, the HS algorithm is applied to search for the optimum solution. The purpose of this research is to maximize the extraterrestrial radiation on the collector surface for a specific period. The sun's position is predicted by the proposed model at different times, and then solar radiation is obtained on various inclined planes with different orientations in each city. The performance of the HS method is compared with that of the ergodic method and other optimization algorithms. The results demonstrate that the tilt angle should be changed once a month, and the best orientation is usually due south in the selected cities. In addition, the HS algorithm is a practical and reliable alternative for estimating the optimum tilt angle and optimum azimuth angle of PV panels.

Keywords: harmony search meta-heuristic algorithm; solar radiation; photovoltaic; tilt angle; orientation

1. Introduction

Environmental pressure and increasing energy costs escalation create a series of fundamental dilemmas in electric power production. Like many developing countries, the growing consumption of energy fuels bolsters the Chinese economy, but increasing the use also exposes potential disruptions in supply [1]. Thus, renewable energy plays an increasingly important part in the future Chinese power system, supplanting all or part of conventional energy sources. Solar energy deserves attention because of its clean, non-polluting, and sustainable use, and other advantages [2]. It is an ideal energy that fundamentally solves the energy crisis and environmental problems [3]. Solar radiation varies with geographic latitude, season and time of a day due to the various solar positions [4]. To maximize the collection of solar radiation, a PV panel should be installed at the appropriate tilt angle and orientation under various circumstances [5].

Recently, many investigators have searched for the optimum tilt angle (β_{opt}) and optimum azimuth angle (γ_{opt}) of solar collectors. Dixit [6] used the artificial neural network (ANN) estimator taking the H_g, ϕ and E_L of the site as inputs, to estimate the optimum tilt angle almost instantaneously while testing. Gopinathan [7] represented the optimum slope and the azimuth angles with the anisotropic model for South Africa, calculating the daily radiation at various slopes and orientations, thus obtaining β_{opt} and γ_{opt}. The past few decades have seen an increased interest in general-purpose "black-box" optimization algorithms that have drawn inspiration from optimization processes that occur in nature in large part [8–14]. In Ref. [15], an approach combining the orthogonal array experimental technique and an ant direction hybrid differential evolution algorithm (ADHDEOA) was presented for determining the tilt angle for PV modules. In Ref. [16], the tilt angle was changed from $-20°$ to $90°$ in a step size of $0.1°$, and the corresponding value of maximum global solar radiation for a specific period is defined as the optimal tilt angle. In Ref. [17], a particle-swarm optimization method with nonlinear time-varying evolution (PSO-NTVE) was proposed, by which the tilt angle of PV modules were determined for Taiwan. The calculation can be formulated as an optimization problem for maximizing the output electrical energy of the modules. From previous applications, the defect of particle swarm optimization (PSO) prematurity makes it easy to fall into a local optimum; thus, it is necessary to select other algorithms to attempt this research.

In 2001, Geem [18] developed a new harmony search (HS) meta-heuristic algorithm that was conceptualized using the musical process of searching for a perfect state of harmony. The word "heuristic" refers to "solution by trial and error method", and the word "meta" refers to "high level". The harmony in music is analogous to the optimization solution vector, and the musician's improvisations are analogous to local and global search schemes in optimization techniques [19]. The characteristics of the HS algorithm include the following three aspects: (1) There is no need for initial values of the decision variables; (2) HS uses a stochastic random search, which requires the harmony memory (HM) without any derivative information; and, (3) When compared with meta-heuristic optimization algorithms, it demands fewer mathematical requirements and can be applied to a wider range. Sandgren [20] and Wu and Chow [21] applied the algorithm to analyze pressure vessel design to minimize the total cost of the material. The HS algorithm was used to solve a weld beam design optimization problem in [19], which was compared with previous solutions. In [22], the HS algorithm was used to determine the near-global optimal initial weights when training the model. Since the initial values of design variables are not required in HS, it is regarded as the best solution in terms of results, and has the fewest limitations on the range of applications.

Traditional methods, such as the ergodic method, are too slow to find the best result in a relatively complex mathematical model. Hence, the HS algorithm is proposed to search for the optimum values because of the advantages mentioned. In this paper, first, the ergodic method is adopted to obtain β_{opt} and γ_{opt} by calculating the extraterrestrial solar radiation at various tilt angles and azimuth angles. The ergodic results are used as standard values for comparison. Next, the calculation of β_{opt} and γ_{opt} is formulated as an optimization problem. Then, the HS method is employed to determine the optimal angles based on the established objective function and constraints. Finally, the HS and PSO results are compared with reference values obtained from the ergodic algorithm through the most widely used statistical methods. Comparisons show that the HS is an accurate and reliable method for the PV module to determine the tilt angle and orientation.

2. Mathematical Model

To obtain the maximum solar radiation, it is necessary to design the tilt angle and orientation of a PV panel. The proper mathematical model related to the latitude of the station and the Julian day is a basis of calculating β_{opt} and γ_{opt}. There are various methods to classify climates in China [23–25]. In this paper, several methods are applied to solar energy collection optimization on the tilted surfaces for six stations (Sanya, Shanghai, Zhengzhou, Harbin, Mohe, and Lhasa) in different climate zones (tropical zone [TZ], subtropical zone [SZ], warm temperate zone [WTZ], mid temperate zone [MTZ], cold temperate zone [CTZ], and Tibetan plateau zone [TPZ]) in China [26]. General layout of the six major climates across China is shown in Figure 1. General information about the selected six typical stations is shown in Table 1.

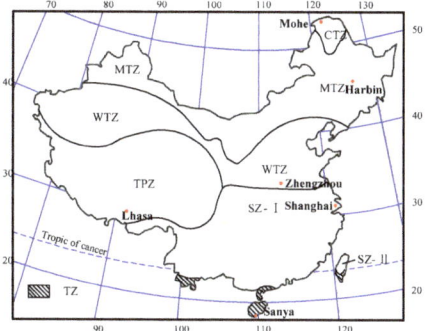

Figure 1. General layout of the six major climates across China. TZ = tropical zone; SZ = subtropical zone; WTZ = warm temperate zone; MTZ = mid temperate zone; CTZ = cold temperate zone; TPZ = Tibetan plateau zone.

Table 1. General information about selected six typical stations.

Climate	Location	Latitude (Φ)	Longitude (E)	Elevation (m)
TZ	Sanya	18°14′	109°31′	5.9
SZ	Shanghai	31°24′	121°29′	6
WTZ	Zhengzhou	34°43′	113°39′	110.4
MTZ	Harbin	45°45′	126°46′	142.3
CTZ	Mohe	53°28′	122°31′	433
TPZ	Lhasa	29°40′	91°08′	3648.7

TZ = tropical zone; SZ = subtropical zone; WTZ = warm temperate zone; MTZ = mid temperate zone; CTZ = cold temperate zone; TPZ = Tibetan plateau zone.

2.1. Julian Day (JD)

The Julian day or Julian day number (JDN) is the number of days that have elapsed since the first day of a year, 1 January. 31 December has a JDN of 365, except when it has a Julian number of 366 in a leap year [27–29].

2.2. Solar Declination (δ)

The declination of the sun is the angle between the line joining the centres of the earth, the sun, and the equatorial plane. In Equation (1), the declination of the sun is determined by the day of the year using the following formula [30]:

$$\delta = 23.45 \sin(2\pi(284+n)/365), \tag{1}$$

where n is the JDN.

2.3. Angle Incidence (θ)

The angle of incidence of the direct solar radiation on the tilt surface, θ, can be calculated by (2) [31]:

$$\begin{aligned}\cos\theta = \; & \sin\delta\sin\phi\cos\beta - \sin\delta\cos\phi\sin\beta\cos\gamma \\ & + \cos\delta\cos\phi\cos\beta\cos\omega \\ & + \cos\delta\sin\phi\sin\beta\cos\gamma\cos\omega \\ & + \cos\delta\sin\beta\sin\gamma\sin\omega\end{aligned}, \tag{2}$$

where ϕ is the latitude of the site, β is the tilt angle of PV panel, γ is the azimuth angle, and ω is the hour angle, which shifts with the sun movement.

2.4. Sunrise and Sunset Hour Angle

The solar altitude angle is zero at the sunrise and sunset. To find the sunrise (or sunset) hour angle ω_r (ω_s) on a south-facing ($\gamma = 0$) tilt surface, one can use the following formulas [32]:

$$\omega_1 = \cos^{-1}(-A/(2B)), \qquad (3)$$

$$\omega_2 = \cos^{-1}(-\tan\phi\tan\delta). \qquad (4)$$

Furthermore, the sunrise (or sunset) hour angle ω_r (ω_s) for the inclined surface is given by [31]:

$$\omega_r = \max(-\omega_1, -\omega_2), \qquad (5)$$

$$\omega_s = \min(\omega_1, \omega_2), \qquad (6)$$

with

$$A = 2(\sin\delta\sin\phi\cos\beta - \sin\delta\cos\phi\sin\beta\cos\gamma) \times (\cos\delta\cos\phi\cos\beta + \cos\delta\sin\phi\sin\beta\cos\gamma), \qquad (7)$$

and

$$B = (\cos\delta\cos\phi\cos\beta + \cos\delta\sin\phi\sin\beta\cos\gamma)^2 \times (\cos\delta\sin\beta\cos\gamma)^2, \qquad (8)$$

where A and B are calculated by Equation (2), which were used to obtain the values of sunrise (or sunset) hour angle in Equations (3)–(6).

2.5. Extraterrestrial Solar Radiation

We use I^* to denote extraterrestrial solar radiation, which is calculated by:

$$I^* = \frac{24 \times 3600}{\pi} I_c \times (1 + 0.033\cos\frac{360n}{365}) \times (\cos\phi\cos\delta\sin\omega_s + \frac{2\pi\omega_s}{360}\sin\phi\sin\delta), \qquad (9)$$

where I_c is the solar constant (=1367 W/m^2).
The extraterrestrial solar radiation on a tilted surface for a day (I_d) is calculated from (10):

$$I_d = \frac{24.3600}{2\pi}\int_{\omega_r}^{\omega_s} I^* \cos\theta d\omega. \qquad (10)$$

The monthly mean daily extraterrestrial solar radiation on a tilted surface (I_m) can be calculated from Equation (11):

$$I_m = \sum_{i=n_1}^{n_2} I_{di}/(n_2 - n_1 + 1), \qquad (11)$$

where n_1 and n_2 are the JD number of the first day and the last day of a month, respectively. I_{di} is the extraterrestrial solar radiation on a tilted surface of the day, which has the JD number of i.

In this paper, the ergodic method employs Equations (1)–(11) to calculate the monthly mean daily extraterrestrial solar radiation for six different stations. The tilt angle is changed from 0° to 90° with a step size of 0.1°, while the azimuth angle is changed from 0° to 360° in the step of 0.1°, and the corresponding value of maximum extraterrestrial solar radiation for a specific period are defined as the optimal angle. Note that the results are considered to be standard values, serving as a reference group in the following research.

3. Object System by Using HS Theory

3.1. Objective Function

In general, an objective function may be any functional relation that an investigator selects to reflect the relative desirability of groupings [33]. According to the "radiation-maximized" demand

and the mathematical model of solar obit and position, this paper establishes the object system of decision-making. Equation (12) reflects the optimization problem, in which the tilt angle and azimuth angle are determined with a maximum of I_m:

$$\max(I_m(\beta, \gamma)). \tag{12}$$

3.2. Constraints

1. Tilt angles

$$\beta_{\min} \leq \beta \leq \beta_{\max}, \tag{13}$$

where $\beta_{\min}$ and $\beta_{\max}$ are the lower and the upper values of β. In this optimization problem, $\beta_{\min}$ and $\beta_{\max}$ are set as $0°$ and $90°$, respectively.

2. Azimuth angles

$$\gamma_{\min} \leq \gamma \leq \gamma_{\max}, \tag{14}$$

where $\gamma_{\min}$ and $\gamma_{\max}$ are the lower and the upper value of γ. In this optimization problem, $\gamma_{\min}$ and $\gamma_{\max}$ are set as $0°$ and $360°$, respectively.

3.3. HS Searching Procedure

The HS algorithm was conceptualized using the musical process of searching for a perfect state of harmony. Musical performances seek pleasing harmony (a perfect state), as determined by an aesthetic standard, just as the optimization process seeks to find a global solution (a perfect state) as determined by an objective function. The pitch of each musical instrument determines the aesthetic quality, just as the objective function value is determined by the set of values assigned to each decision variable [11]. The searching procedure using HS theory for β_{opt} and γ_{opt} [13] is as follows:

Step 1: Initialize the optimization problem and algorithm parameters. First, the optimization problem is specified as Equations (12)–(14). The HS algorithm parameters required to solve the optimization problem (i.e., Equations (12)–(14) are also set in this step: harmony memory size (number of solution vectors in harmony memory, HMS) = 6, harmony memory considering rate (HMCR) = 0.9, pitch adjusting rate (PAR) = (0.4, 0.9), and termination criterion (maximum number of searches) = 2000.

Step 2: Initialize the harmony memory (HM). The solution vectors in the HM matrix are generated randomly and sorted by the values of the objective function. The HM is given by (15):

$$HM = \begin{bmatrix} \beta^1 & \beta^2 & \cdots & \beta^{HMS} \\ \gamma^1 & \gamma^2 & \cdots & \gamma^{HMS} \end{bmatrix}^T. \tag{15}$$

Step 3: Improvise a new harmony from the HM. A new harmony vector, (β', γ') is generated from the HM based on memory considerations, pitch adjustments, and randomization. For example, the value of the first design variable (β') for the new vector can be chosen from any value in the specified HM range $(\beta^1 - \beta^{HMS})$. Values of the other design variables (γ') can be chosen in the same way. Here, the algorithm chooses the new value with HMCR = 0.9:

$$\beta' \leftarrow \begin{cases} \beta' \in \{\beta^1, \beta^2, \ldots, \beta^{HMS}\}, \text{with probability HMCR} \\ \beta' \in (\beta_{\min}, \beta_{\min}), \text{with probability } (1 - HMCR) \end{cases}, \tag{16}$$

$$\gamma' \leftarrow \begin{cases} \gamma' \in \{\gamma^1, \gamma^2, \ldots, \gamma^{HMS}\}, \text{with probability HMCR} \\ \gamma' \in (\gamma_{\min}, \gamma_{\min}), \text{with probability } (1 - HMCR) \end{cases}. \tag{17}$$

The pitch adjusting process is performed until a value is chosen from the HM. PAR of 0.4 indicates that it is possible to choose a neighboring value using 40% × HMCR. The value (1-PAR) sets the rate of doing nothing.

$$\beta', \gamma' \leftarrow \begin{cases} \text{Pitch adjusting decision for} \\ \quad \text{Yes with probability PAR} \\ \quad \text{No with probability } (1 - PAR) \end{cases}. \quad (18)$$

If the pitch adjustment decision for β', γ' is "Yes", and β', γ' are assumed to be β' (k), γ' (k), the kth element in β and γ, the pitch-adjusted values of β (k), (k) are:

$$\beta' \leftarrow \beta' + \alpha, \quad (19)$$

$$\gamma' \leftarrow \gamma' + \alpha, \quad (20)$$

where $\alpha = bw \times u(-1, 1)$, $bw \in (0.0001, 1)$, which is an arbitrary distance bandwidth for the continuous design variable, and $u(-1,1)$ is a uniform distribution between -1 and 1.

Step 4: Update the HM. If the objective function value of new harmony vector is larger than the worst harmony in the HM, include the new harmony in the HM and exclude the existing worst harmony from the HM. Then, sort the HM by the objective function value.

Step 5: Repeat Steps 3 and 4 until the termination criterion is met. In this step, the termination of a computation process is allowed before the final conclusion, in accordance with specified termination criterion (maximum number of searches = 2000). If not, repeat Steps 3 and 4.

4. Statistical Methods

To assess the accuracy and suitability of the models, numerous statistical methods are used to compare the reference optimum tilt angle obtained by the ergodic method and the optimum tilt angle calculated by the PSO and HS algorithms. PSO is a stochastic technique for exploring the search space for optimization [34]. It emulates the social behavior of a natural swarm. With the term "swarm", a population of individuals (particles) is indicated. Each swarm particle follows a path in the solution space. The motion of each particle is governed by a list of constraints based on ecology (i.e., natural swarm behavior) [35]. In the present study, PSO is used to compare the solution quality with the HS algorithm. The model equations are evaluated using mean percentage error (MPE), mean absolute percentage error (MAPE), mean absolute bias error (MABE), and root mean square error (RMSE), which are defined as below [36–42]:

$$MPE = \frac{1}{N} \sum_{i=1}^{N} \left(\frac{\beta_{ic} - \beta_{im}}{\beta_{im}} \cdot 100 \right), \quad (21)$$

$$MAPE = \frac{1}{N} \sum_{i=1}^{N} \left(\left| \frac{\beta_{ic} - \beta_{im}}{\beta_{im}} \right| \cdot 100 \right), \quad (22)$$

$$MABE = \sum_{i=1}^{N} |\beta_{ic} - \beta_{im}| / N, \quad (23)$$

$$RMSE = \sqrt{\sum_{i=1}^{N} (\beta_{ic} - \beta_{im})^2 / N}, \quad (24)$$

where β_{ic} and β_{im} are the ith calculated and standard optimum tilt angles, respectively; N is the total number of observations; and β_{ca} and β_{ma} are the average of the calculated and standard values, respectively.

5. Results and Discussions

The optimum tilt angles and orientations of PV modules for six stations in different climates are established using the HS method, accounting for the climate and latitude of each site. The six Chinese cities selected for study are Sanya, Shanghai, Zhengzhou, Harbin, Mohe, and Lhasa. In practical applications, PV modules do not operate under a standard condition. During the sun's movements, the solar irradiance accumulates each hour angle with the corresponding extraterrestrial radiation.

Based on Equations (1)–(11), the monthly mean daily extraterrestrial radiation data were determined accordingly; Figure 2 displays the results from January to December in Shanghai. Taking the result of March as an example, in Figure 2c, I_{dm} on a tilted surface varies from 0 to over 35 MJ/m^2. Apart from south-facing orientations, the solar radiation decreases gradually with an increasing incline angle from horizontal to vertical surfaces. The maximum value is observed at the inclined angles between 20° and 40°, with the azimuth angles from 150° to 210°. The figure also indicates that I_d is quite symmetrical with respect to due south ($\gamma = 180°$). As the result of the sun path in Shanghai, the sun is visible for most of the day throughout the year; therefore, solar collection is greatest among all of the surfaces. Since the east-facing surfaces and west-facing surfaces face the sun in the morning and afternoon, respectively, the solar radiations at these two orientations are very close to each other. Due to the shortest period facing the direct sun, the north-facing planes receive the least solar radiation.

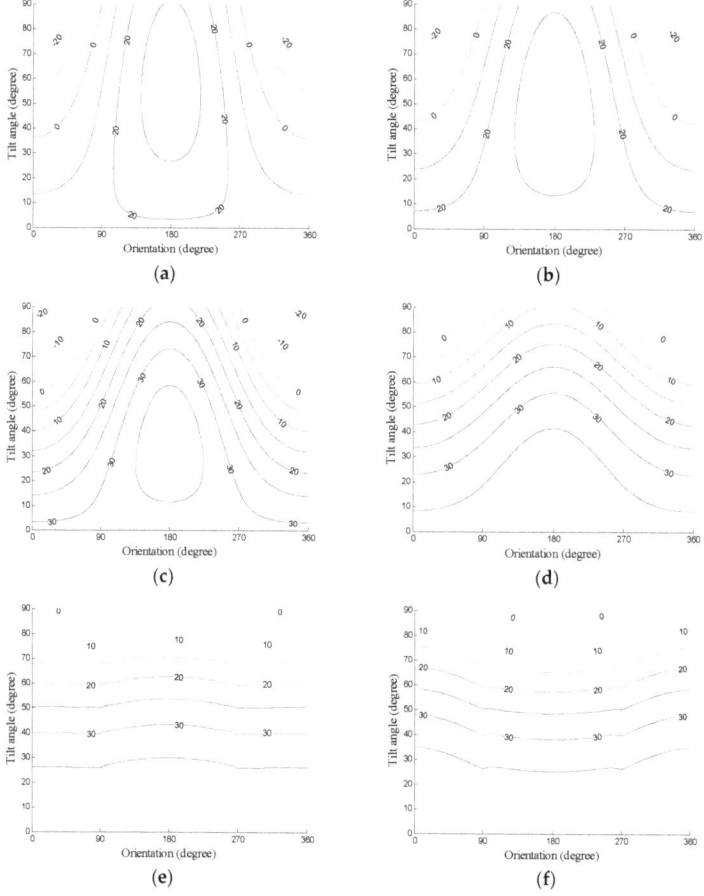

Figure 2. *Cont.*

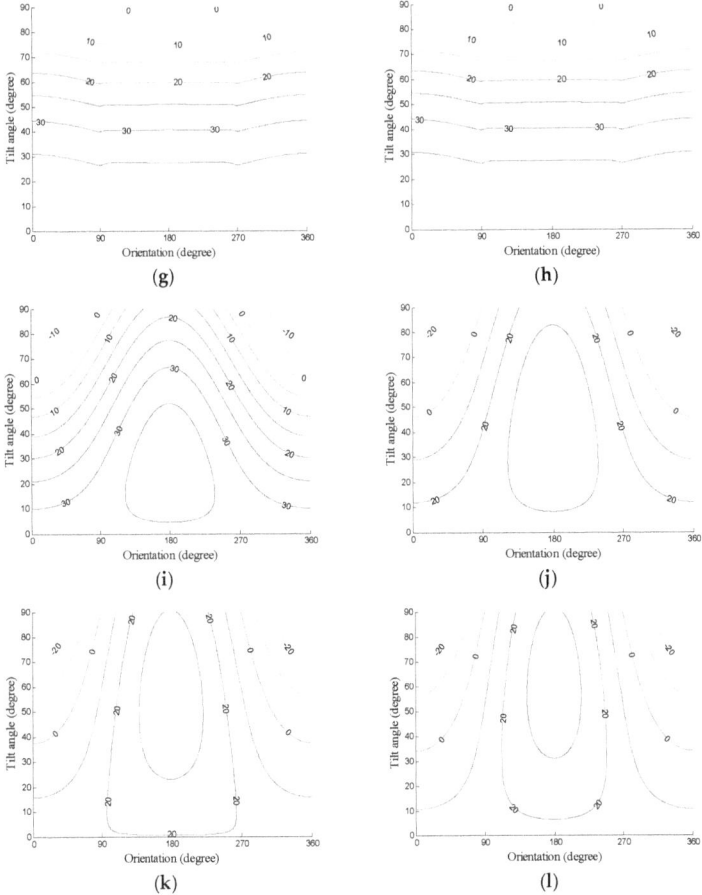

Figure 2. Monthly daily extraterrestrial solar radiation for various tilted angles and orientations in Shanghai: (**a**) Jan; (**b**) Feb; (**c**) Mar; (**d**) Apr; (**e**) May; (**f**) Jun; (**g**) Jul; (**h**) Aug; (**i**) Sep; (**j**) Oct; (**k**) Nov; and, (**l**) Dec.

From Figure 2, the optimum azimuth angles are 0° (360°) or 180° in every month. Therefore, when referring to the optimum results, the azimuth angle data are simplified to the sign in the discussion below: a positive sign means due south, and a negative sign means due north. Table 2 demonstrates the values of β_{opt} obtained using the ergodic method, the PSO and the HS algorithm. In the ergodic results, β_{opt} values from Sanya, Shanghai, Zhengzhou, Harbin, Mohe, and Lhasa ranges from −18.1°(Jun) to 49.9°(Dec), from −7.6°(Jun) to 61.4°(Dec), from 5.5°(Jun) to 64.3°(Dec), from 12.6°(Jun) to 73.7°(Dec), from 16.6°(Jun) to 80.0°(Dec), and from −8.9°(Jun) to 59.9°(Dec), respectively. The values of β_{opt} are negative from May to July. The results also show that γ_{opt} values obtained with the HS method are exactly the same as those obtained with the PSO method and ergodic method, and that the trend of β_{opt} values in each month is roughly the same.

Table 2. Optimum tilt angles of each station obtained by ergodic, particle swarm optimization (PSO), and harmony search (HS) methods.

Station	Method	Month											
		Jan	Feb	Mar	Apr	May	Jun	Jul	Aug	Sep	Oct	Nov	Dec
Sanya (TZ)	Ergodic	47.2	37.7	21.8	3.6	−11.7	−18.1	−15.2	3.1	15.1	32.9	45.0	49.9
	PSO	47.45	37.17	22.23	3.74	−11.88	−18.14	−14.67	3.28	15.60	32.66	45.25	49.78
	HS	47.24	37.73	21.90	3.67	−11.76	−18.12	−15.22	3.15	15.11	32.95	45.08	49.86
Shanghai (SZ)	Ergodic	59.0	50.1	34.9	16.6	5.3	−7.6	3.8	11.1	28.3	45.6	57.0	61.4
	PSO	58.19	49.17	33.89	15.22	5.49	−8.32	4.02	11.35	28.17	44.33	56.26	60.51
	HS	58.99	50.18	35.00	16.56	5.33	−7.58	3.81	11.07	28.26	45.66	56.96	61.40
Zhengzhou (WTZ)	Ergodic	61.9	53.2	38.2	19.8	5.5	6.4	14.2	31.6	48.8	59.9	64.3	
	PSO	61.95	53.06	38.36	19.37	7.74	5.45	6.31	13.89	31.52	48.80	59.83	64.19
	HS	61.93	53.20	38.29	19.80	8.01	5.52	6.42	14.13	31.58	48.85	59.88	64.28
Harbin (MTZ)	Ergodic	71.5	63.5	49.1	30.8	16.5	12.6	14.0	24.7	42.6	59.3	69.7	73.7
	PSO	71.59	63.46	49.83	30.67	16.15	12.82	13.96	23.67	42.69	59.14	69.41	73.58
	HS	71.52	63.65	49.23	30.49	16.56	12.62	13.96	24.73	42.58	59.29	69.47	73.68
Mohe (CTZ)	Ergodic	77.9	70.5	56.8	38.7	23.0	16.6	19.2	32.7	50.3	66.5	76.3	80.0
	PSO	77.69	70.33	56.90	38.47	22.12	16.57	19.20	31.20	50.60	66.30	76.48	80.04
	HS	77.90	70.51	56.86	38.57	23.06	16.59	19.16	32.76	50.33	66.47	76.34	80.00
Lhasa (TPZ)	Ergodic	57.5	48.6	33.3	14.9	3.8	−8.9	−5.7	9.5	26.5	44.0	55.4	59.9
	PSO	57.56	48.73	33.14	14.97	3.58	−8.62	−5.64	8.99	26.48	44.18	55.31	59.98
	HS	57.46	48.54	33.28	14.92	3.88	−8.91	−5.69	9.55	26.53	43.99	55.44	59.89

Table 3 compares the optimum tilt angles and monthly total extraterrestrial solar radiation of Shanghai using the ergodic, PSO, HS, and some traditional methods. The results indicate that radiation on $β_{opt}$ obtained with the HS method is clearly higher than that obtained using the PSO method and methods, taking $β$ to be equal to $Φ$ and $Φ ± 15°$. Hence, the tilt angle obtained using $Φ$ for the PV modules should not be adopted for different places to obtain the maximum overall solar energy. However, in Table 3, the results of the HS method are closest to the standard values. These data show that the HS method can be used as a substitute for the ergodic method.

Table 3. Comparison of results of Shanghai using several methods.

Month	Ergodic Method		PSO Method		HS Method		$β_{opt} = Φ$	$β_{opt} = Φ + 15(°)$	$β_{opt} = Φ − 15(°)$
	$β_{opt}(°)$	I(MJ/m²)	$β_{opt}(°)$	I(MJ/m²)	$β_{opt}(°)$	I(MJ/m²)			
Jan	59.0	1100.86	58.19	1100.76	58.99	1100.86	975.89	1074.49	810.79
Feb	50.1	1046.94	49.17	1046.80	50.18	1046.94	991.57	1044.74	870.83
Mar	34.9	1182.78	33.89	1182.58	35	1182.78	1180.54	1159.16	1121.48
Apr	16.6	1156.53	15.22	1156.21	16.56	1156.53	1118.21	1004.55	1156.52
May	5.3	1213.51	5.49	1213.50	5.33	1213.51	1070.68	887.24	1184.17
Jun	−7.6	1180.48	−8.32	1180.38	−7.58	1180.48	982.83	780.58	1122.34
Jul	3.8	1215.11	4.02	1215.10	3.81	1215.11	1042.77	845.05	1173.14
Aug	11.1	1205.31	11.35	1205.29	11.07	1205.31	1129.09	983.04	1199.82
Sep	28.3	1146.41	28.17	1146.41	28.26	1146.41	1144.69	1089.48	1122.01
Oct	45.6	1169.21	44.33	1168.94	45.66	1169.21	1133.67	1169.09	1020.99
Nov	57.0	1083.85	56.26	1083.77	56.96	1083.85	977.85	1065.52	823.54
Dec	61.4	1078.14	60.51	1078.01	61.4	1078.14	933.68	1041.40	762.34

For further comparison, the statistical errors of PSO and HS for six typical climatic stations in China are shown in Table 4, obtained using Equations (21)–(24). According to MPE, MAPE, MABE, and RMSE, the errors of the HS model are lower than those of the PSO method. More specifically, the computational results show that HS errors are about 1 order of magnitude smaller than the PSO errors. The MPE of the HS method is between −0.0450% and 0.4376%, the MAPE ranges from 0.1105% to 0.4510%, the maximum of MABE is 0.0867°, and the minimum of MABE is 0.0317°. The RMSE value, which measures the accuracy of estimation, stays well between 0.0385° and 0.1279°. The data in Table 4 starkly illustrate that the monthly optimum tilt angle data obtained from the HS method are more accurate and reliable than that from PSO. The results indicate that the HS method is applicable to the optimization problem of the tilt angle and orientation for PV modules.

Table 4. Statistical indicators (mean percentage error (MPE), mean absolute percentage error (MAPE), mean absolute bias error (MABE), root mean square error (RMSE)) of PSO and HS at six different climatic stations.

Methods	Statistical Indicators	Sanya	Shanghai	Zhengzhou	Harbin	Mohe	Lhasa
PSO	MPE	0.9968	0.0560	−0.8521	−0.3617	−0.7451	−1.2242
	MAPE	1.9739	3.4607	0.9353	0.9567	0.9215	1.4549
	MABE	0.2825	0.7117	0.1458	0.2742	0.3200	0.1550
	RMSE	0.3266	0.8206	0.1894	0.4005	0.5264	0.2018
HS	MPE	0.4376	0.0337	0.0472	−0.0450	0.0022	0.2163
	MAPE	0.4510	0.2008	0.1507	0.2405	0.1105	0.2943
	MABE	0.0475	0.0383	0.0292	0.0867	0.0392	0.0317
	RMSE	0.0539	0.0476	0.0393	0.1279	0.0524	0.0385

6. Conclusions

The tilt angle and orientation play an important role in maximizing the solar radiation collected by a PV panel. In this paper, the ergodic method is applied to the mathematical model of extraterrestrial solar radiation to determine the monthly optimum tilt angles and azimuth angles in six Chinese cities with different climates. The results of the ergodic method serve as a reference group. Then, the calculation of the optimum angles based on HS theory is presented. By comparing the results of the HS and PSO methods with standard results, the following conclusions can be drawn:

1. In most cases, the best orientation is due south (optimum azimuth angle, 180°) in the selected cities. Except when the azimuth angle equals 180°, the extraterrestrial solar radiation decreases as the tilt angle increases.
2. The optimum tilt angle increases during the winter months and reaches a maximum in December for all of the stations. To enhance the energy collected by the panel, if possible, the tilt angle should be changed once a month.
3. According to MPE, MAPE, MABE, and RMSE, errors with the HS method are less than those with PSO. Moreover, the extraterrestrial solar radiation of HS is larger than that of PSO. The application of HS performs better in the search for β_{opt}.
4. The proposed approach, the HS method, provides an accurate and simple alternative to the ergodic method. The experimental results of the HS method are very close to the standard values.

Acknowledgments: The research is supported by National Natural Science Foundation of China (Program No. 51507052), the Fundamental Research Funds for the Central Universities (Program No. 2015B02714), Jiangsu Key Laboratory of Smart Grid Technology and Equipment, and Science and Technology Project of SGCC (Research and application of key technologies for operation and control of urban smart photovoltaic energy storage charging tower under "source-grid-load" interactive environment). The authors also thank the China Meteorological Administration.

Author Contributions: Mian Guo conceived and designed the experiments and wrote the paper; Haixiang Zang and Shengyu Gao performed the experiments; Tingji Chen, Jing Xiao and Lexiang Cheng analyzed the data; Zhinong Wei and Guoqiang Sun contributed reagents/materials/analysis tools.

Conflicts of Interest: The authors declare no conflict of interest.

References

1. Zhao, H.R.; Guo, S. External benefit evaluation of renewable energy power in china for sustainability. *Sustainability* **2015**, *7*, 4783–4805. [CrossRef]
2. Järvelä, M.; Valkealahti, S. Ideal operation of a photovoltaic power plant equipped with an energy storage system on electricity market. *Appl. Sci.* **2017**, *7*, 749. [CrossRef]
3. Liu, L.-Q.; Wang, Z.-X.; Zhang, H.-Q.; Xue, Y.-C. Solar energy development in china—A review. *Renew. Sustain. Energy Rev.* **2010**, *14*, 301–311. [CrossRef]
4. Li, D.; Lam, T.; Chu, V. Relationship between the total solar radiation on tilted surfaces and the sunshine hours in hong kong. *Sol. Energy* **2008**, *82*, 1220–1228. [CrossRef]

5. Samani, P.; Mendes, A.; Leal, V.; Correia, N. Pre-fabricated, environmentally friendly and energy self-sufficient single-family house in kenya. *J. Clean. Prod.* **2017**, *142*, 2100–2113. [CrossRef]
6. Dixit, T.V.; Yadav, A.; Gupta, S. Optimization of pv array inclination in india using ann estimator: Method comparison study. *Sadhana-Acad. Proc. Eng. Sci.* **2015**, *40*, 1457–1472. [CrossRef]
7. Gopinathan, K.K.; Mallehe, N.B.; Mpholo, M.I. A study on the intercepted insolation as a function of slope and azimuth of the surface. *Energy* **2007**, *32*, 213–220. [CrossRef]
8. Wolpert, D.H.; Macready, W.G. No free lunch theorems for optimization. *IEEE Trans. Evolut. Comput.* **1997**, *1*, 67–82. [CrossRef]
9. Simons, K.T.; Kooperberg, C.; Huang, E.; Baker, D. Assembly of protein tertiary structures from fragments with similar local sequences using simulated annealing and bayesian scoring functions. *J. Mol. Biol.* **1997**, *268*, 209–225. [CrossRef] [PubMed]
10. Dupanloup, I.; Schneider, S.; Excoffier, L. A simulated annealing approach to define the genetic structure of populations. *Mol. Ecol.* **2002**, *11*, 2571–2581. [CrossRef] [PubMed]
11. Konak, A.; Coit, D.W.; Smith, A.E. Multi-objective optimization using genetic algorithms: A tutorial. *Reliab. Eng. Syst. Saf.* **2006**, *91*, 992–1007. [CrossRef]
12. Rudolph, G. Convergence analysis of canonical genetic algorithms. *IEEE Trans. Neural Netw.* **1994**, *5*, 96–101. [CrossRef] [PubMed]
13. Yaow-Ming, C.; Chien-Hsing, L.; Hsu-Chin, W. Calculation of the optimum installation angle for fixed solar-cell panels based on the genetic algorithm and the simulated-annealing method. *IEEE Trans. Energy Convers.* **2005**, *20*, 467–473.
14. Deb, K.; Pratap, A.; Agarwal, S.; Meyarivan, T. A fast and elitist multiobjective genetic algorithm: Nsga-ii. *IEEE Trans. Evolut. Comput.* **2002**, *6*, 182–197. [CrossRef]
15. Chang, Y.-P. An ant direction hybrid differential evolution algorithm in determining the tilt angle for photovoltaic modules. *Expert Syst. Appl.* **2010**, *37*, 5415–5422. [CrossRef]
16. Zang, H.; Guo, M.; Wei, Z.; Sun, G. Determination of the optimal tilt angle of solar collectors for different climates of china. *Sustainability* **2016**, *8*, 654. [CrossRef]
17. Chang, Y.P. Optimal tilt angles for photovoltaic modules using pso method with nonlinear time-varying evolution. *Energy* **2010**, *35*, 1954–1963. [CrossRef]
18. Geem, Z.W.; Kim, J.H.; Loganathan, G.V. A new heuristic optimization algorithm: Harmony search. *Simulation* **2001**, *76*, 60–68. [CrossRef]
19. Lee, K.S.; Geem, Z.W. A new meta-heuristic algorithm for continuous engineering optimization: Harmony search theory and practice. *Comput. Methods Appl. Mech. Eng.* **2005**, *194*, 3902–3933. [CrossRef]
20. Sandgren, E. Nonlinear integer and discrete programming in mechanical design optimization. *J. Mech. Des.* **1990**, *112*, 223–229. [CrossRef]
21. Wu, S.-J.; Chow, P.-T. Genetic algorithms for nonlinear mixed discrete-integer optimization problems via meta-genetic parameter optimization. *Eng. Optim.* **1995**, *24*, 137–159. [CrossRef]
22. Lee, A.; Geem, Z.W.; Suh, K.-D. Determination of optimal initial weights of an artificial neural network by using the harmony search algorithm: Application to breakwater armor stones. *Appl. Sci.* **2016**, *6*, 164. [CrossRef]
23. Lam, J.C.; Wan, K.K.; Yang, L. Solar radiation modelling using anns for different climates in china. *Energy Convers. Manag.* **2008**, *49*, 1080–1090. [CrossRef]
24. Lau, C.C.; Lam, J.C.; Yang, L. Climate classification and passive solar design implications in china. *Energy Convers. Manag.* **2007**, *48*, 2006–2015. [CrossRef]
25. Yang, L.; Wan, K.K.; Li, D.H.; Lam, J.C. A new method to develop typical weather years in different climates for building energy use studies. *Energy* **2011**, *36*, 6121–6129. [CrossRef]
26. Zang, H.; Xu, Q.; Bian, H. Generation of typical solar radiation data for different climates of china. *Energy* **2012**, *38*, 236–248. [CrossRef]
27. Tang, R.S.; Tong, W. Optimal tilt-angles for solar collectors used in china. *Appl. Energy* **2004**, *79*, 239–248. [CrossRef]
28. Tang, R.S.; Yu, Y.M. Feasibility and optical performance of one axis three positions sun-tracking polar-axis aligned cpcs for photovoltaic applications. *Sol. Energy* **2010**, *84*, 1666–1675. [CrossRef]
29. Sakonidou, E.P.; Karapantsios, T.D.; Balouktsis, A.I.; Chassapis, D. Modeling of the optimum tilt of a solar chimney for maximum air flow (vol 82, pg 80, 2008). *Sol. Energy* **2012**, *86*, 809. [CrossRef]

30. Huang, B.J.; Sun, F.S. Feasibility study of one axis three positions tracking solar pv with low concentration ratio reflector. *Energy Convers. Manag.* **2007**, *48*, 1273–1280. [CrossRef]
31. Duffie, J.A.; Beckman, W.A. *Solar Engineering of Thermal Processes*; John Wiley and Sons: New York, NY, USA, 1980.
32. Schaap, A.B.; Veltkamp, W.B. *Solar Engineering of Thermal Processes*, 2nd ed.; Elsevier Science Publishers: New York, NY, USA, 1993; Volume 51, p. 521.
33. Ward, J.H. Hierarchical grouping to optimize an objective function. *J. Am. Stat. Assoc.* **1963**, *58*, 236–244. [CrossRef]
34. Kennedy, J.; Eberhart, R. *Particle Swarm Optimization*; IEEE: New York, NY, USA, 1995; pp. 1942–1948.
35. Fulginei, F.R.; Salvini, A. Comparative analysis between modern heuristics and hybrid algorithms. *Compel-Int. J. Comp. Math. Electr. Electron. Eng.* **2007**, *26*, 259–268. [CrossRef]
36. Li, H.; Ma, W.; Lian, Y.; Wang, X. Estimating daily global solar radiation by day of year in china. *Appl. Energy* **2010**, *87*, 3011–3017. [CrossRef]
37. Jiang, Y. Generation of typical meteorological year for different climates of china. *Energy* **2010**, *35*, 1946–1953. [CrossRef]
38. Bulut, H.; Büyükalaca, O. Simple model for the generation of daily global solar-radiation data in turkey. *Appl. Energy* **2007**, *84*, 477–491. [CrossRef]
39. Ampratwum, D.B.; Dorvlo, A.S. Estimation of solar radiation from the number of sunshine hours. *Appl. Energy* **1999**, *63*, 161–167. [CrossRef]
40. Menges, H.O.; Ertekin, C.; Sonmete, M.H. Evaluation of global solar radiation models for konya, turkey. *Energy Convers. Manag.* **2006**, *47*, 3149–3173. [CrossRef]
41. Jiang, Y. Computation of monthly mean daily global solar radiation in china using artificial neural networks and comparison with other empirical models. *Energy* **2009**, *34*, 1276–1283. [CrossRef]
42. Jin, Z.; Yezheng, W.; Gang, Y. General formula for estimation of monthly average daily global solar radiation in china. *Energy Convers. Manag.* **2005**, *46*, 257–268. [CrossRef]

© 2017 by the authors. Licensee MDPI, Basel, Switzerland. This article is an open access article distributed under the terms and conditions of the Creative Commons Attribution (CC BY) license (http://creativecommons.org/licenses/by/4.0/).

Article

ANN Sizing Procedure for the Day-Ahead Output Power Forecast of a PV Plant

Francesco Grimaccia, Sonia Leva, Marco Mussetta and Emanuele Ogliari *

Department of Energy, Politecnico di Milano, 20156 Milano, Italy; francesco.grimaccia@polimi.it (F.G.); sonia.leva@polimi.it (S.L.); marco.mussetta@polimi.it (M.M.)
* Correspondence: emanuelegiovanni.ogliari@polimi.it; Tel.: +39-2399-8524

Academic Editor: Allen Barnett
Received: 4 May 2017; Accepted: 12 June 2017; Published: 15 June 2017

Abstract: Since the beginning of this century, the share of renewables in Europe's total power capacity has almost doubled, becoming the largest source of its electricity production. In 2015 alone, photovoltaic (PV) energy generation rose with a rate of more than 5%; nowadays, Germany, Italy, and Spain account together for almost 70% of total European PV generation. In this context, the so-called *day-ahead electricity market* represents a key trading platform, where prices and exchanged hourly quantities of energy are defined 24 h in advance. Thus, PV power forecasting in an *open energy market* can greatly benefit from machine learning techniques. In this study, the authors propose a general procedure to set up the main parameters of hybrid artificial neural networks (ANNs) in terms of the number of neurons, layout, and multiple trials. Numerical simulations on real PV plant data are performed, to assess the effectiveness of the proposed methodology on the basis of statistical indexes, and to optimize the forecasting network performance.

Keywords: artificial neural network; day-ahead forecast; ensemble methods

1. Introduction

The recent increase of renewable energy sources (RES) quota in power systems is facing new technical challenges, involving also the overall efficiency of the electrical grid [1]. Therefore, several techniques able to forecast the output power of RES facilities have been developed, in particular addressing the inherent variability of parameters related to solar radiation and atmospheric weather, which directly affect photovoltaic (PV) systems' power output [2]. Robust predictive tools based on the measured historical data of a specific PV plant can provide advantages in its operation policy, in order to avoid or minimize excess production, and to benefit from the available incentives for producing electricity from RES [3]. Moreover, the increasing penetration of accurate forecasting models, constrained by the market and incentives, can implicitly facilitate the alignment of producers' decisions with energy policy and emissions targets.

In particular, these predictive models are aimed at finding out the relationship between the numerical weather prediction and the forecasted power output [4]. However, any forecasting activity of natural phenomena is characterized by a high degree of non-linearity. To address this complexity, artificial neural networks (ANNs) with a multilayer perceptron (MLP) architecture are often used. These are a class of machine learning (ML) techniques used to solve specific kinds of problems, such as: pattern recognition, function approximation, control, and forecasting [5–7].

ANNs are known as a very flexible tool, and several layouts have been developed to solve different tasks [8]. In particular, ML forecasting methods can be used on an extended dataset of historical measurements, providing an advantage in reducing prediction errors with respect to other statistical and physical forecasting models [9,10]. An additional advantage of ML techniques is their ability to handle missing data and to solve problems with a high degree of complexity.

The first breakthroughs in using ML technologies in the solar power forecasting field were pioneered more than a decade ago [11]. A comprehensive review of solar power forecast approaches is given in [12]; this paper provides a systematic discussion comparing and contrasting many works on applying artificial neural networks to the PV forecasting problem. ML techniques are demonstrated to be successfully applied for this purpose, provided that suitable historical patterns are available. In addition, the potential benefits of hybrid techniques, such as those in [13,14], are also thoroughly discussed.

Nowadays, in the case of a particular PV plant's power output, the common training data are the historical measurements of power production, and the meteorological parameters related to the specific plant's location, which include temperature, global horizontal irradiance, and cloud cover above the facility. Additional forecasted variables from numerical weather predictions can also be considered, such as wind speed, humidity, and pressure.

Recently, novel forecasting models have been implemented by adding an estimate of the clear sky radiation to the series of historical local weather data, in order to hybridize the meteorological forecasts with physical data, as reported in [15].

Moreover, the effectiveness of using ensemble methods, i.e., running simultaneously a number of parallel neural networks and finally averaging their results, was demonstrated in [16], thus giving additional advantages also to modern parallel computing techniques. The same paper also shows the response of the network to changes in training set size and meteorological conditions.

In this paper, the authors present a procedure to set up the main characteristics of the network using a physical hybrid method (PHANN) to perform the day-ahead PV power forecast in view of the electricity market. The procedure outlined here can be adopted to set up the best settings of the network in terms of the number of layers, neurons, and trials. Once all the optimum settings have been identified after this procedure, it is possible to perform the day-ahead forecast with the PHANN method using different sets for training and validation. The test data set will be made up of the 24 hourly PV power values forecasted one day ahead.

The paper is structured as follows: Section 2 provides an overview of the considered neural network architectures. Section 3 presents the methodology implemented to compare different ANN structures and dimensions, and Section 4 will propose some metrics aimed at evaluating each configuration's suitability in terms of error performance and statistical behavior. Section 5 presents the considered case study, which is used to test the proposed methodology in terms of the number of layers, neurons and the size of the ensemble. Specific simulations and numerical results are provided in Sections 6–8, and final remarks are reported in Section 9.

2. Artificial Neural Networks

An ANN basically consists of a number of neurons, grouped in different layers, which receive input from all of the neurons in the preceding layer. Each neuron performs a simple nonlinear operation, and the learning capacity is finite due to the number of neurons and links existing among them. The number of neurons in the first input layer is equal to the number of input data, and the number of neurons in the output layer corresponds to the number of output values. Hidden layers are in between, and their neurons can vary in number.

Choosing the amount of neurons in the layers and the number of layers depends on several factors, as well as the kind problem to be addressed [17]. There are several proposed models to define an optimal network [18], but these methods are not applicable to every area of study. Indeed, the success of a network depends heavily on the experience of the creator [19], and many attempts must be made to reach satisfactory result. Moreover, some limits are imposed by the computational burden. In fact, the more complex the network, the longer the time (hours) needed for data processing.

This paper describes a heuristic process, by means of a statistical indicators comparison, in order to optimize the settings of a Feed Forward Neural Network (FFNN) for PV forecasting based on weather predictions. The search for the best layout is constrained by computational limitations. For the above mentioned FFNN, the most effective configuration based on the minimum absolute mean error (normalized or weighted) index has been studied. Since the future weather conditions are unknown at the time of prediction, it is impossible to determine a priori which precise network layout will provide the best PV power production forecast.

It must be pointed out that, during the FFNN training step, both the input (i.e., the historical weather forecasts) and the target (i.e., the historical PV plant output power) datasets employed are just one representative couple of the virtually infinite couples of input and target presented in the historical weather data pool.

Furthermore, for any given dataset, each network configuration presents an additional element of randomness. In fact, for a specific network layout, the training process is initialized by means of randomly chosen weights and biases; therefore, for the same forecast, the performance is non-unique.

Hence, to set up the main settings, it is useful to feed the FFNN with the same data that is representative of the majority of the possible mutations incurring in the dataset. A one-year dataset represents a valid data sample, as it spans over periods of different weather conditions. Still, if the network deals with a climatically unusual day forecast, a worsening of the performance should be expected.

3. Methodology

3.1. Early Assumptions

As one of the goals of this work is to study how the number of neurons in the ANN's layers affects the forecast, and to consequently determine the optimal setup (that is, the solution that minimizes the errors), we will modify only the network settings, while keeping the same database. Since this is a stochastic method applied on the available dataset, an average performance must be evaluated, and often the "best solution" will not be unique.

In fact, two networks will provide different outputs even if equally configured to follow the same training process. This is due to the random initialization of weights and biases, which is peculiar for each training run. Therefore, comparing two networks with different settings only by looking at their output cannot generally lead to comprehensive answers. A rigorous approach towards stochastic elements is specifically needed, especially in the comparison between various network layouts and settings, when searching for the best possible configuration. Hence, FFNN has been analyzed by studying the output of the network as a function of the number of neurons and their layout. In each study, the neuron activation function was kept unchanged, and we used the tan-sigmoid (*"tansig"* in Matlab Neural Network Toolbox™, R2016a, MathWorks®, Natick, MA, USA [20]), which has the following expression, where x is the generic input and y the output of the neuron:

$$y = \frac{2}{1 + e^{-2x}} - 1. \tag{1}$$

Here, two "similar networks" are defined as two networks with the same topology (number of neurons and their layout), but with different weights and biases. Therefore, statistic tools are used to infer more tangible conclusions, giving more strength to the final settings adopted by the method.

3.2. Data Set Definition

In order to perform the forecast, historical data are provided to the Neural Network Toolbox™ [20] in Matlab software, R2016a, MathWorks®, Natick, MA, USA, and grouped in three clusters: "training", "validation", and "test". Each group fulfills three specific tasks:

(1) The *training set* includes the samples employed to train the network. It should contain enough different examples (days) to make the network able to generalize its learning.
(2) The *validation set* contains additional samples (i.e., days not already included in the training set) used by the network to check and validate the training process.

(3) The *test set* is the dataset corresponding to the days actually forecasted by the network.

3.3. Artificial Neural Network Size

The ANN sizing problem primarily means to choose how many neurons are in the layers, and their layout. The neurons (or units) in the input and output layers are fixed by the problem to be solved, while characterizing the neurons in the hidden layers is controversial, since the hidden neurons are regarded as the processing neurons in the network. In addition, having a small number of hidden layers might increase the speed of the training process, whereas a large number of hidden layers could make it longer. Furthermore, the nonlinearity capability of an ANN increases with the number of layers; i.e., the more complicated (nonlinear) the input/output relationship is, the more layers an ANN will need to model it. In addition, the greater the number of neurons per layer, the more accurately can ANNs identify input/output relationships. In order to guarantee a sufficient generalization capability for the ANN, the number of neurons is reasonably bound to the training set size. Though, from the literature, only some general conditions and intuitive rules could be inferred.

In Widrow's review [21], it is stated that the number of weights N_w existing within a layered ANN is bound to the number of patterns N_p in the training set and outputs N_o by the following condition:

$$N_p \gg N_w/N_o. \tag{2}$$

Condition (2) means that a given ANN should have an amount of neurons that is sufficiently smaller than the training samples. In fact, if the degrees of freedom (i.e., N_w) are larger than the constraints associated with the desired response function (i.e., N_o and N_p), the training procedure will be unable to completely constrain the weights in the network.

In the mid-1960s, Cover studied the capacity of an FFNN with an arbitrary number of layers and a single output element [22,23]. In the same line, Baum and Haussler [21] addressed the question of the net size, which provides a valid generalization stating, with a desired accuracy of 90%, that at least 10 times as many training examples as there are weights in the ANN should be considered. In [24], the authors also provide theoretical lower and upper bounds on the sample size as a function of the net size, such that a valid generalization can be expected. However, they limited their study under the assumption that the node functions are linear threshold functions (or at least Boolean valued), leaving open the problem for classes of real valued functions (such as sigmoid functions), and multiple hidden layers networks.

To characterize the number of neurons in the layers, there are different rules suggested in the literature, starting from two incremental algorithms ([25] and the reference therein), passing through a trial and error procedure [26], to some thumb rules. For example, Chow et al. [27] applied—in the same forecasting context of this work—a thumb rule found in [28] in the "backpropagation architecture—standard connections" section. There, it is stated that the default number of hidden neurons N_n for a three layer network is computed with the following formula:

$$N_n = \frac{(i+o)}{2} + \sqrt{N_p} \tag{3}$$

where i is the number of inputs; and o is the number of outputs. For more layers, it is stated that N_n should be divided by the number of hidden layers.

Finally, there is a symmetrical technique from the already mentioned tiling algorithm [29]. In the first, a small number of neurons is selected. Then, they are increased gradually. The symmetrical approach is referred to as the "pruning" algorithm [30]. This method, starting from a large network, carries on a gradual removal of the neurons by erasing the less significant units, and it requires in advance the largest size of the network [31].

Although the above cited approaches have a general value, in this paper the authors present a procedure to select the best settings of the neural network in terms of the number of layers, neurons, and trials, with the specific application of the physical hybrid method (PHANN) performing the day-ahead PV power forecast in view of the electricity market.

4. Evaluation Indexes

With respect to our specific application in PV power forecasting, although several evaluation indexes with different meanings are adopted in the literature, their trends are usually highly correlated, as they are calculated on the common basis of the hourly error e_h:

$$e_h = P_{m,h} - P_{p,h} \qquad (4)$$

which is the difference between the measured value of the output power $P_{m,h}$ and the forecasted one $P_{p,h}$ in the same hour h. Therefore, we now consider the normalized mean absolute error $NMAE_\%$ as a reference for evaluating the performance of the forecasts:

$$NMAE_\% = \frac{1}{N \cdot C} \sum_{i=1}^{N} \left| P_{m,h} - P_{p,h} \right| \cdot 100. \qquad (5)$$

The normalized mean absolute error $NMAE_\%$ is based on the net capacity of the plant C. N is the number of time samples (hours) considered in the evaluated period (i.e., $N = 24$ in a daily error basis calculation).

This index was calculated for the output "test set" of each individual network's performance. As the $NMAE_\%$ value is a random variable, its trend could be analyzed by means of the theory of parametric estimation. Therefore, when looking for the settings which on average minimize $NMAE_\%$ values, the FFNN will be analyzed as a function of:

(1) a single layer number of neurons;
(2) a double layer number of neurons; and
(3) trials in the ensemble forecast.

To proceed with this analysis, these steps are followed:

(1) one or more ANN parameters are kept constant;
(2) the free parameter to be inspected (i.e., number of neurons within a layer) is varied within a specific range; and
(3) the $NMAE_\%$ values of similar networks are calculated.

Under appropriate conditions, a range of values of the $NMAE_\%$ can be calculated, assuming that the unknown mean of all of the possible $NMAE_\%$ values, with those ANN settings, is within that interval of confidence.

NMAE Statistical Distributions and Confidence Limits

In order to study $NMAE_\%$ behavior with respect to a specific ANN setting, this setting is assigned a starting constant value, and n_t forecasts are performed. For each i-th forecast (trial), the $NMAE_\%$ is calculated. From a statistical point of view, the group of these $NMAE_\%$ values represents a sample of the endless population of all of the possible $NMAE_\%$ values related to the forecast performed with those specific ANN settings. After the same parameter has been changed, further n_t forecasts are performed, and the associated $NMAE_\%$ values are calculated. This procedure is repeated until the maximum value of the ANN parameter is reached. As this value could be as great as possible, a reasonable threshold is set. Intuitively, a higher accuracy could be obtained by increasing the number of forecasts for each network, but the highly time-consuming process should be compensated for by a much more striking performance, otherwise it is not worthy. Therefore, a tradeoff between the computational burden and the expected accuracy is defined. Considering the group of $NMAE_\%$ values belonging to the same test set obtained by the ANN with those given "p" settings, the relative sample mean $\overline{NMAE}_p$ is an estimator of all of the possible $NMAE_\%$ values, and it is defined as:

$$\overline{NMAE}_p = \frac{1}{n_t} \sum_{i=1}^{n_t} NMAE_{i,p} \qquad (6)$$

where $NMAE_{i,p}$ is the $NMAE_\%$ calculated for the i-th trial performed by the ANN with the p-th value of a given setting. In our case, the distribution of the $NMAE_{i,p}$ population is unknown. However, the sample mean can be calculated (3), as well as the sample variance S_p^2:

$$S_p^2 = \frac{1}{n_t - 1} \sum_{i=1}^{n_t} (NMAE_{i,p} - \overline{NMAE}_p)^2. \qquad (7)$$

The sample standard deviation, S_p, is:

$$S_p = \sqrt{\sum_{i=1}^{n_t} \frac{(NMAE_{i,p} - \overline{NMAE}_p)^2}{n_t - 1}}. \qquad (8)$$

Any statistic estimating the value of a parameter is an "estimator" or a "point estimator" [32]. It is often impossible to know if the point estimator is correct, because it is merely an estimation of the actual value (which is, in our case, impossible to find, unless after performing an endless number of forecasts). For this reason, we construct confidence limits CI helping the estimation of the unknown population mean μ, which are defined as the sample mean with a margin of error ME:

$$CI = \overline{NMAE}_p \pm ME. \qquad (9)$$

The ME is set according to:

(1) how confident we want to be with our assessment;
(2) the sample standard deviation S; and
(3) how large our sample size is.

In this way, we are able to define the probability α that the mean of the population μ has to fall outside our CI, which is split into the two tails of the probability density curve for the t-Student curve. In this way, the ME is defined as:

$$ME = t_{\frac{\alpha}{2}} \left(\frac{S_p}{\sqrt{n_t}} \right). \qquad (10)$$

For example, when $\alpha = 0.05$ (which implies that CI = 95%), the critical value $t_{\alpha/2}$ is set by the relative t-Student distribution selected according to the degree of freedom equal to $n_t - 1$. Now that the distribution has been estimated, it is possible to define appropriate confidence intervals in which the mean μ of the population can be included, according to the sample mean $\overline{NMAE}_p$.

5. Case Study

This study was lead on three different PV plants located in the northern part of Italy, each with different features. Nevertheless, it is important to highlight that this procedure has general validity, and can be easily extended to other PV plant locations with different data set availability [33,34].

The available hourly datasets considered here cover one year of measurements of the PV output power. The historical weather forecasts for the next day, which have been employed to train the network, are delivered daily by a weather service at 11 a.m. The historical hourly database of some weather parameters, such as the ambient temperature, global horizontal solar radiation, wind, etc., used to train the neural network. In addition, the deterministic global solar radiation under clear sky conditions is provided to the network for obtaining an hybrid method [7]. The number and type of these input parameters have been defined in [33,34], where the comparison of the neural approach versus physical methods has been presented. As reported in the cited paper, physical approaches based only on temperature and irradiance are outperformed by neural networks considering an extended range of meteorological parameters.

In order to properly train the network, the three groups of data (training, validation, and test) should be statistically consistent: for every physical parameter describing the data point, they should contain almost the same mean value, variability, and range. The samples' grouping is realized as follows: 70% for training, 15% for validation, and 15% for testing. The samples composing the three groups are randomly chosen from the whole dataset. The data are first divided into 365 blocks (one block per day); then, 255 days are assigned to training, 55 days to validation, and 55 days are left for the test. It has to be noted that each group has the same average number of daylight hours.

Finally, for each p-th configuration of the network, we calculated a sufficiently large number of forecasts (i.e., 100) related to 55 days, and covering all of the possible meteorological conditions (test set). For each 55-day forecast, the corresponding $NMAE_\%$ was evaluated. The $NMAE_\%$'s sample mean, its sample variance, and the width of its confidence interval for a given degree of confidence were also calculated as previously described. As the sample variance is different for each network configuration, also the related confidence intervals will have different amplitudes.

6. Neurons in a Single Layer ANN

The method illustrated above has been applied to define the number of neurons in a single layer FFNN. First, we examined some networks fed with just one layer between the input and the output. We considered two different types of training, as well as a number of neurons varying between 20 and 160, with a resolution equal to 2 up to 80 neurons, and equal to 4 onwards. Then, for every different topology, 100 training runs were carried out, and as many outputs and relative $NMAE_\%$s were calculated. Afterwards, we calculated the average and the variance of this parameter, and we chose intervals of 95% confidence for the sample mean. Both methods used the Levenberg–Marquardt (LM) algorithm, together with an early stopping procedure. The "LM fast" method adopts the default setting, while the second one has been changed in order to assume a slower convergence towards the solution. This ensures higher protection from overlearning, as suggested in the user's guide [20], together with the default values already set in the Neural Network Toolbox™. More in detail, the parameters for the two "LM" methods are shown in Table 1, where ω is the initial convergence speed, δ is the speed increase and Φ the speed decrease when the convergence speed is inadequate.

Table 1. Levenberg–Marquardt (LM) training algorithm settings.

Settings	LM Fast (Default)	LM Slow
ω	10^{-3}	1
δ	10	2
Φ	10^{-1}	-0.5

The results can be observed in the Figures 1 and 2, where the mean $NMAE_\%$ trend as a function of the neurons in the single hidden layer is shown. The error function reaches a minimum within a broad interval of neurons, after which it seems to increase again. This slight growth is not evident in the slow convergence algorithm ("LM slow", Figure 2); however, we may expect that this will not decrease further, as the error has reached a minimum region. In fact, while an exact point for the minimum of the sample mean can be determined, (the red points in Figures 1 and 2), this value is included in an interval of confidence. If such a range matches other minima (which are not necessarily adjacent), it would not be possible to exactly determine which one of the two points actually represents the absolute minimum (the "optimum" value). Two intervals of confidence are called "compatible" if their intersection is not null. The mean points representing intervals compatible with the minimum interval are shown in Figures 1 and 2 (red point), and have been highlighted in yellow. It can be noted that, for the fast convergence algorithm ("LM fast", Figure 1) with an equal number of similar networks, the trend is more variable: the higher variance generates wider intervals. As a consequence, we have a higher number of compatible points, and higher uncertainty in the optimum configuration.

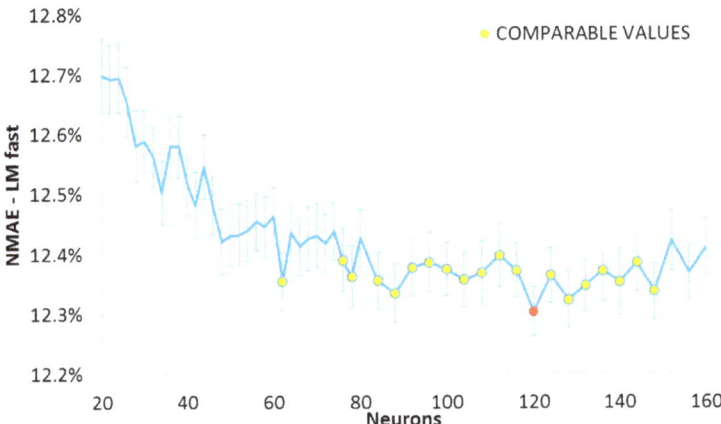

Figure 1. Mean normalized mean absolute error ($NMAE_\%$) as a function of the number of neurons, with Levenberg–Marquardt training algorithm set for a faster convergence. Comparable mean $NMAE_\%$ in yellow, with 95% interval of confidence, with the minimum in red.

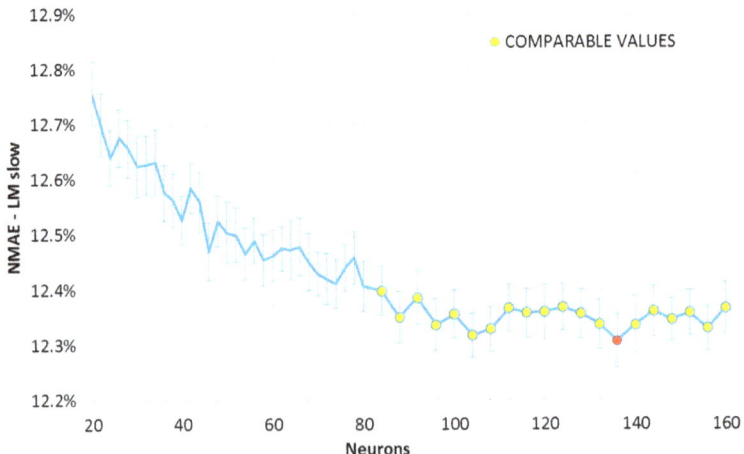

Figure 2. Mean $NMAE_\%$ as a function of the number of neurons, with Levenberg–Marquardt training algorithm set for a slower convergence. Comparable mean $NMAE_\%$ in yellow, with 95% interval of confidence, with the minimum in red.

It is useful to recall here that such intervals contain the real mean value within a 95% degree of confidence. It is then plausible to assume that some of these points are outliers (consider for example the value for 62 neurons in Figure 1). Indeed, the error variation within the optimal range is shown to be minimal in these graphs. Besides, it is better to consider also the error from the ensemble forecasting method as it is described in the following section. Looking at Figures 1 and 2, a suitable number of neurons for a single layer FFNN is between 100 and 120, with a training-validation dataset composition as explained in Section 4.

The proposed approach provides generally good and flexible results on a whole year data basis, including all kinds of meteorological conditions. The daily errors in different weather conditions according to the presented methodology are variable, as already shown in [15]. The forecasting performances (as daily $NMAE_\%$) vary according to the weather conditions, as follows: around 2% for a sunny day; 7% for highly variable cloudy day; and 8% for a typical overcast day. The proposed PHANN approach has been already validated versus other methods for a day-ahead PV power forecasts in [33]: results obtained in a real application scenario, after setting neural network parameters with the here proposed methodology, have proved to reach lower error rates.

7. Number of Trials in the Ensemble Forecast

In order to obtain the lowest error, it is also worthwhile to estimate how many trials should be employed in the ensemble forecast. The improvement is expected to reach an asymptotic value with a growing number of trials; however, by increasing the amount of trials, the calculation time is enlarged. Therefore, there is a threshold gain between the number of trials and the reached improvement.

The procedure of looking for the minimum $NMAE_\%$, along with the amount of trials, is outlined as follows. The training Levenberg–Marquardt algorithms both with fast and slow convergence are adopted, and the ensemble forecast is performed by an ANN with 120 number of neurons. A growing number of trials were analyzed up to a maximum equal to one thousand.

First, the $NMAE_\%$ is calculated for ten independent ensemble forecasts. The term "independent" means that different ensemble forecasts do not have common trials. The n-th ensemble forecast is performed by using a growing quantity of trials, starting from one trial and going up to one hundred trials, in order to infer the global trends of the ensemble $NMAE_\%$, while trying to avoid any possible random influence due to a few number of cases.

Figures 3 and 4 show the results of this analysis for the LM fast and slow training algorithms, respectively. They represent the ten $NMAE_\%$ error curves as a function of the average outputs number used for the ensemble method (grey lines). The mean $NMAE_\%$ of the ten ensemble forecasts is depicted in red; the mean $NMAE_\%$ of one thousand forecasts is the upper constant dashed green line; and the $NMAE_\%$ of the ensemble forecast made by one thousand trials is the lower constant dash-dotted blue line. The red line represents a "trend index", and it rapidly tends to an asymptotic value that can be considered "stable".

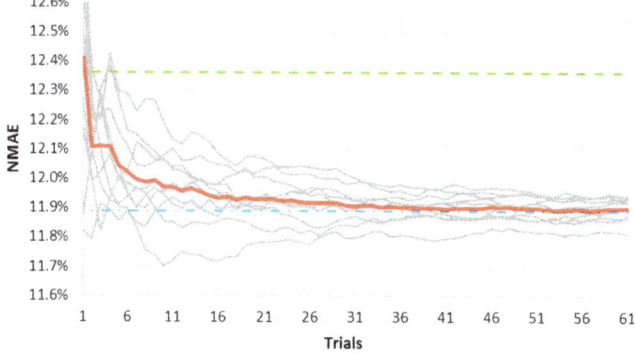

Figure 3. $NMAE_\%$ of ten random ensemble (grey lines) as a function of the growing number of trials. The artificial neural network (ANN) has 120 neurons and an "LM fast" training algorithm. The mean of the ten ensemble forecasts $\overline{NMAE_p}$ is in red, the mean $NMAE_\%$ of one thousand forecasts is the dashed green line, and the ensemble $NMAE_\%$ of one thousand forecasts is the dash-dotted blue line.

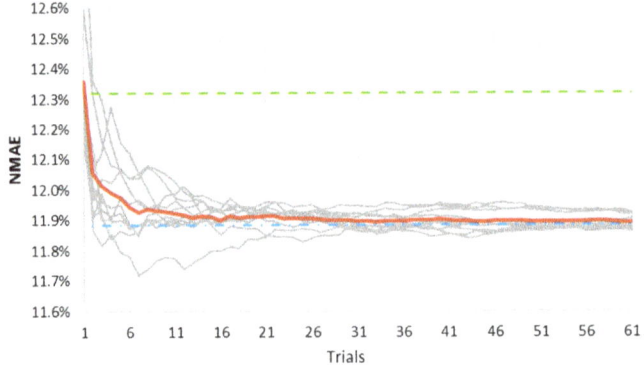

Figure 4. $NMAE_\%$ of ten random ensemble (grey lines) as a function of the growing number of trials. The artificial neural network (ANN) has 120 neurons and a "Levenberg–Marquardt (LM) slow" training algorithm. The mean of the ten ensemble forecasts $\overline{NMAE_p}$ is in red, the mean of one thousand forecasts is the dashed green line, and the ensemble $NMAE_\%$ of one thousand forecasts is the dash-dotted blue line.

It must be pointed out that as far as the number of trials gets close to one thousand, the red and the blue line will tend to coincide. However, the graphs have an upper limit equal to 100 trials in the x axis, as we consider ten independent ensembles from one thousand different forecasts.

When the slow convergence algorithm is used, such stability is reached for a smaller number of trials, as shown in Figure 4. Furthermore, the curves' variability here is lower, reflecting the lower error variance of the output. Even the advantage initially offered by the fast convergence algorithm—that is, the short calculation time—seems now to be compensated for by the need of a lower number of outputs in the slow convergence algorithm. A larger number of forecasts in the ensemble method would slightly decrease $NMAE_\%$, which bears the brunt of a much higher computational burden.

8. Number of Neurons in a Dual Layer ANN

In this work, networks with two hidden layers between the input and the output have also been analyzed. This network topology has two degrees of freedom related to the optimum number of neurons. Therefore, different criteria can be adopted, such as keeping the number of neurons in one layer constant while varying them in the other, or keeping a constant ratio of neurons between the layers, etc. Several simulations have been performed, but hereafter only the Levenberg–Marquardt training algorithm with a slow convergence setting is considered.

Figure 5 shows the mean $NMAE_\%$ trends of the dual layer networks as a function of the neurons of the first layer, for a total of 50 forecasts for each layout. From 20 to 120 neurons were employed, with an increasing rate of 20 neurons. In Figure 6, the neurons of the second layer were fixed at 25%, 50%, and 75% of the first layer neurons, rounded up, and originating in this way three different curves. The best result is comparable to the one reached by the single layer networks. However, here it is possible, by analyzing the gradient of the orange curve, to have small margins of improvement while the number of neurons is increasing. Similarly to the analysis carried out for a single layer network, the $NMAE_\%$ is calculated for the ensemble of the forecasts, and it is shown together with the mean error of the single outputs. Also in this case, the shape of the curves suggests looking for the optimum layout towards a higher number of neurons. Moreover, the smallest value of the error for the ensemble output is slightly less than the one obtained by the single layer networks. Therefore, the dual layer networks seem to provide better performance.

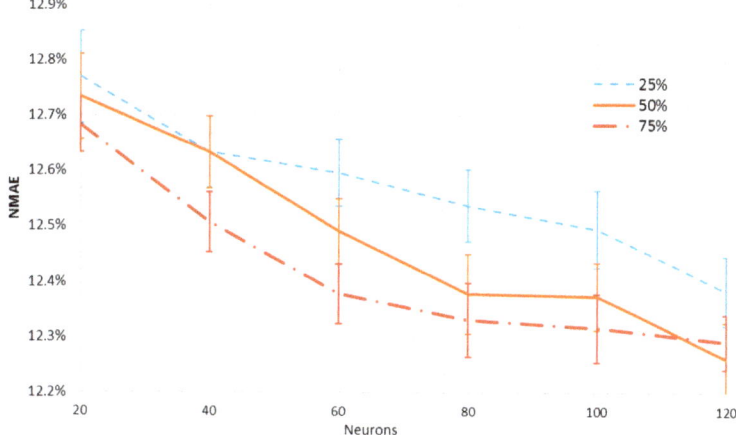

Figure 5. Mean $NMAE_\%$ of 50 different forecasts as a function of the hidden layers' sizes, with a constant ratio of neurons.

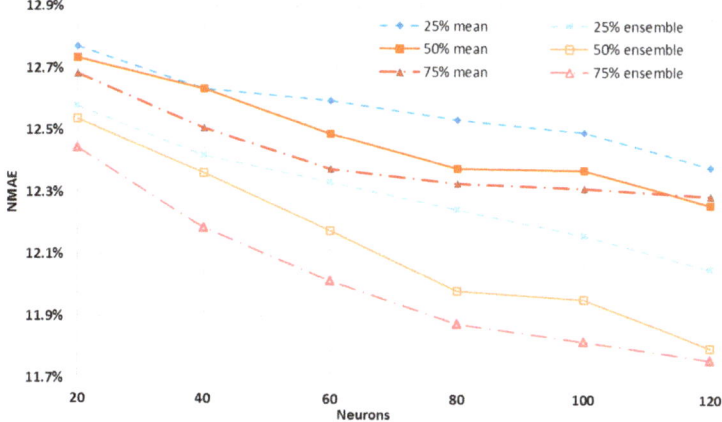

Figure 6. Comparison between the mean $NMAE_\%$ and the ensemble $NMAE_\%$ of 50 trials as a function of the hidden layers' sizes, kept with a constant ratio of neurons.

Although the dual layer networks seem to provide slightly better performance, single layer networks seem to be more appropriate for this kind of forecast. In fact, dual layer networks with larger numbers of neurons need remarkably higher calculation times, and this might represent a great drawback for their employment. As an example, the training of only one single layer network with 120 neurons required an average of 50 s, while the training of a unique dual layer network with 120 and 90 neurons required nearly 350 s on a quad-core Intel®, Santa Clara, CA, USA, Core™ i7-2640M CPU, with an operating frequency of 2.8 GHz, and 8 GB ram.

9. Conclusions

This work has presented a detailed analysis of a method which has been defined for the setting up of an ANN in terms of number of trials in the ensemble forecast, the number of neurons, and their layout. This method has been employed for a day-ahead forecast with a recently developed PHANN approach.

On the basis of the proposed method, the settings minimizing the mean $NMAE_\%$ of the 24 h ahead PV power forecast with a one-year historical dataset are: an ensemble size of ten trials, and a number of 120 neurons in a single layer ANN configuration. The above outlined method is meant to be adopted for setting the most suitable ANN parameters in view of the day-ahead forecast of any PV plant's output power.

Author Contributions: In this research activity, all of the authors were involved in the data analysis and preprocessing phase, the simulation, the results analysis and discussion, and the manuscript's preparation. All of the authors have approved the submitted manuscript. S.L. and M.M. conceived and designed the experiments; E.O. performed the experiments; E.O. and M.M. analyzed the data; F.G. contributed analysis tools; all the authors equally contributed to the writing of the paper.

Conflicts of Interest: The authors declare no conflict of interest.

References

1. Brenna, M.; Dolara, A.; Foiadelli, F.; Lazaroiu, G.C.; Leva, S. Transient analysis of large scale PV systems with floating DC section. *Energies* **2012**, *5*, 3736–3752. [CrossRef]
2. Paulescu, M.; Paulescu, E.; Gravila, P.; Badescu, V. *Weather Modeling and Forecasting of PV Systems Operation*; Green Energy Technology; Springer: London, UK, 2013.
3. Omar, M.; Dolara, A.; Magistrati, G.; Mussetta, M.; Ogliari, E.; Viola, F. Day-ahead forecasting for photovoltaic power using artificial neural networks ensembles. In Proceedings of the 2016 IEEE International Conference on Renewable Energy Research and Applications (ICRERA), Birmingham, UK, 20–23 November 2016; pp. 1152–1157.
4. Cali, Ü. *Grid and Market Integration of Large-Scale Wind Farms Using Advanced Wind Power Forecasting: Technical and Energy Economic Aspects*; Kassel University Press: Kassel, Germany, 2011.
5. Gardner, M.; Dorling, S. Artificial neural networks (the multilayer perceptron)—A review of applications in the atmospheric sciences. *Atmos. Environ.* **1998**, *32*, 2627–2636. [CrossRef]
6. Lippmann, R. An introduction to computing with neural nets. *IEEE ASSP Mag.* **1987**, *4*, 4–22. [CrossRef]
7. Gandelli, A.; Grimaccia, F.; Leva, S.; Mussetta, M.; Ogliari, E. Hybrid model analysis and validation for PV energy production forecasting. In Proceedings of the International Joint Conference on Neural Networks, Beijing, China, 6–11 July 2014.
8. Nelson, M.M.; Illingworth, W.T. *A Practical Guide to Neural Nets*; Addison-Wesley: Reading, MA, USA, 1991; Volume 1.
9. Pelland, S.; Remund, J.; Kleissl, J.; Oozeki, T.; De Brabandere, K. Photovoltaic and Solar Forecasting: State of the Art. *Int. Energy Agency Photovolt. Power Syst. Program. Rep.* **2013**, *14*, 1–40.
10. Jayaweera, D. *Smart Power Systems and Renewable Energy System Integration*; Springer International Publishing: Cham, Switzerland, 2016; Volume 57.
11. Kalogirou, S.A. *Artificial Intelligence in Energy and Renewable Energy Systems*; Nova Science Publishers: Commack, NY, USA, 2006.
12. Raza, M.Q.; Nadarajah, M.; Ekanayake, C. On recent advances in PV output power forecast. *Sol. Energy* **2016**, *136*, 125–144. [CrossRef]
13. Li, G.; Shi, J.; Zhou, J. Bayesian adaptive combination of short-term wind speed forecasts from neural network models. *Renew. Energy* **2011**, *36*, 352–359. [CrossRef]
14. Quan, D.M.; Ogliari, E.; Grimaccia, F.; Leva, S.; Mussetta, M. Hybrid model for hourly forecast of photovoltaic and wind power. In Proceedings of the IEEE International Conference on Fuzzy Systems, Hyderabad, India, 7–10 July 2013.
15. Dolara, A.; Grimaccia, F.; Leva, S.; Mussetta, M.; Ogliari, E. A physical hybrid artificial neural network for short term forecasting of PV plant power output. *Energies* **2015**, *8*, 1138–1153. [CrossRef]
16. Leva, S.; Dolara, A.; Grimaccia, F.; Mussetta, M.; Ogliari, E. Analysis and validation of 24 hours ahead neural network forecasting of photovoltaic output power. *Math. Comput. Simul.* **2017**, *131*, 88–100. [CrossRef]
17. Ripley, B.D. Statistical Ideas for Selecting Network Architectures. In *Neural Networks: Artificial Intelligence and Industrial Applications, Proceedings of the Third Annual SNN Symposium on Neural Networks, Nijmegen, The Netherlands, 14–15 September 1995*; Kappen, B., Gielen, S., Eds.; Springer: London, UK; pp. 183–190.

18. Benardos, P.G.; Vosniakos, G.-C. Optimizing feedforward artificial neural network architecture. *Eng. Appl. Artif. Intell.* **2007**, *20*, 365–382. [CrossRef]
19. Basheer, I.; Hajmeer, M. Artificial neural networks: Fundamentals, computing, design, and application. *J. Microbiol. Methods* **2000**, *43*, 3–31. [CrossRef]
20. Beale, M.H.; Hagan, M.T.; Demuth, H.B. *Neural Network Toolbox^{TM} User's Guide*; MathWorks Inc.: Natick, MA, USA, 1992.
21. Widrow, B.; Lehr, M.A. 30 Years of Adaptive Neural Networks: Perceptron, Madaline, and Backpropagation. *Proc. IEEE* **1990**, *78*, 1415–1442. [CrossRef]
22. Cover, T.M. *Geometrical and Statistical Properties of Linear Threshold Devices*; Department of Electrical Engineering, Stanford University: Stanford, CA, USA, 1964.
23. Cover, T.M. Capacity problems for linear machines. *Pattern Recognit.* **1968**, 283–289.
24. Baum, E.B.; Haussler, D. What Size Net Gives Valid Generalization? *Neural Comput.* **1989**, *1*, 151–160. [CrossRef]
25. Hertz, J.; Krogh, A.; Palmer, R.G. *Introduction to the Theory of Neural Computation*; Westview Press: Boulder, CO, USA, 1991; Volume 1.
26. Grossman, T.; Meir, R.; Domany, E. Learning by Choice of Internal Representations. *Complex Syst.* **1989**, *2*, 555–575.
27. Chow, S.K.H.; Lee, E.W.M.; Li, D.H.W. Short-term prediction of photovoltaic energy generation by intelligent approach. *Energy Build.* **2012**, *55*, 660–667. [CrossRef]
28. Frederick, M. *Neuroshell Manual*; Ward Systems Group Inc.: Frederick, MD, USA, 1996; Volume 2.
29. Mezard, M.; Nadal, J.-P. Learning in feedforward layered networks: The tiling algorithm. *J. Phys. A Math. Gen.* **1989**, *22*, 2191–2203. [CrossRef]
30. Castellano, G.; Fanelli, A.M.; Pelillo, M. An iterative pruning algorithm for feedforward neural networks. *IEEE Trans. Neural Netw.* **1997**, *8*, 519–531. [CrossRef] [PubMed]
31. Huang, S.-C.; Huang, Y.-F. Bounds on the number of hidden neurons in multilayer perceptrons. *IEEE Trans. Neural Netw.* **1991**, *2*, 47–55. [CrossRef] [PubMed]
32. Shao, J. Fundamentals of Statistics. In *Mathematical Statistics*; Springer: New York, NY, USA, 2003; pp. 91–160.
33. Ogliari, E.; Dolara, A.; Manzolini, G.; Leva, S. Physical and hybrid methods comparison for the day ahead PV output power forecast. *Renew. Energy* **2017**, *113*, 11–21. [CrossRef]
34. Dolara, A.; Leva, S.; Mussetta, M.; Ogliari, E. PV hourly day-ahead power forecasting in a micro grid context. In Proceedings of the EEEIC 2016—International Conference on Environment and Electrical Engineering, Florence, Italy, 6–8 June 2016.

 © 2017 by the authors. Licensee MDPI, Basel, Switzerland. This article is an open access article distributed under the terms and conditions of the Creative Commons Attribution (CC BY) license (http://creativecommons.org/licenses/by/4.0/).

Article

Novel Genetic Algorithm-Based Energy Management in a Factory Power System Considering Uncertain Photovoltaic Energies

Ying-Yi Hong * and Po-Sheng Yo

Department of Electrical Engineering, Chung Yuan Christian University, Taoyuan 32023, Taiwan; posheng.yo@gmail.com
* Correspondence: yyhong@ee.cycu.edu.tw; Tel.: +886-3-265-1200

Academic Editors: Emanuele Giovanni Carlo Ogliari and Sonia Leva
Received: 25 February 2017; Accepted: 21 April 2017; Published: 26 April 2017

Abstract: The demand response and accommodation of different renewable energy resources are essential factors in a modern smart microgrid. This paper investigates the energy management related to the short-term (24 h) unit commitment and demand response in a factory power system with uncertain photovoltaic power generation. Elastic loads may be activated subject to their operation constraints in a manner determined by the electricity prices while inelastic loads are inflexibly fixed in each hour. The generation of power from photovoltaic arrays is modeled as a Gaussian distribution owing to its uncertainty. This problem is formulated as a stochastic mixed-integer optimization problem and solved using two levels of algorithms: the master level determines the optimal states of the units (e.g., micro-turbine generators) and elastic loads; and the slave level concerns optimal real power scheduling and power purchase/sale from/to the utility, subject to system operating constraints. This paper proposes two novel encoding schemes used in genetic algorithms on the master level; the point estimate method, incorporating the interior point algorithm, is used on the slave level. Various scenarios in a 30-bus factory power system are studied to reveal the applicability of the proposed method.

Keywords: demand response; genetic algorithm; renewable energy; unit commitment; uncertainty

1. Introduction

Unit commitment (UC) for the power utility determines the states (on/off) of each thermal generator over the scheduled time horizon. Recently, the problem of the high penetration of renewables in power systems has become important. Bertsimas et al. proposed a two-stage adaptive robust UC model for the security-constrained UC problem in the presence of nodal net injection uncertainty using Benders decomposition algorithm [1]. Kalantari et al. projected all feasible generation and demand vectors onto the demand space and reformulated the UC within this loadability set [2]. Ignoring the uncertainty, Bakirtzis et al. used a mixed integer linear programming based on different time-resolutions to study the UC problem, considering renewables [3].

On the other hand, demand response (DR) is the change in electric usage by end-use customers from their normal consumption patterns in response to changes in the electricity price associated with the retail market or incentive policies [4–6]. Customer participation in the DR program is a key factor to enhance the reliability in a modern smart grid.

Traditionally, DR has been addressed in the scheduling of household appliances for their energy consumption in response to retail prices [7,8] or the electricity prices in the day ahead power market [9]. Yoon et al. proposed a dynamic demand response controller that changed the set-point temperature to control HVAC loads, depending on electricity retail price, which was published each 15 min, and partially shifted some of this load away from the peak [10]. Tsui and Chan developed a versatile DR optimization framework that used convex programming for the automatic load management of various household appliances in a smart home [11]. Pipattanasomporn et al. proposed an algorithm to manage household loads according to their preset priorities and to guarantee that total household power consumption was below certain levels [12]. Chavali et al. proposed an approximate greedy iterative algorithm to schedule the energy usages of appliances of end-use customers. Each customer in the system will find the optimal start time and operating mode of each appliance in response to varying electricity prices [13].

DR can also be achieved by alternative approaches, such as aggregated load control [14,15], plug-in electric vehicles [16] and frequency-related control [17]. Pourmousavi et al. evaluated the thermostat setpoint control of aggregate electric water heaters for load shifting, and for providing a desired balancing reserve for the utility. This work also assessed the economic benefits of DR for customers considering time-of-use pricing [14]. Salinas et al. considered a third-party's management of the energy consumption of a group of users, and formulated the corresponding load scheduling problem as a constrained multi-objective optimization problem. The optimization objectives were to minimize the cost of consumed energy and to maximize a certain utility, which can be conflicting and non-commensurable [15]. Tan et al. investigated a market in which users have the flexibility to sell back the energy generated by their distributed generators or the energy stored in their plug-in electric vehicles, using a distributed optimization algorithm [16]. Chang-Chien et al. proposed an overall frequency restoration plan that considered the DR and spinning reserve. The DR program was used first to restore declining frequency caused by a large disturbance. The scheduled spinning reserve was then used to raise frequency back to the pre-disturbance level [17].

The concepts of DR, UC and renewable energy as they pertain to a power utility were recently integrated [18–20]. Abdollahi et al. investigated the economically driven and environmentally driven measures of DR programs and proposed a new linearized formulation of the cost-emission-based UC problem that was associated with DR program solved by mixed-integer programming [18]. Zhao et al. addressed DR programs as another reserve resource to mitigate uncertainty in wind power output; they developed a robust optimization approach to derive an optimal UC decision by maximizing total social welfare under the joint worst-case wind power output and DR scenario [19]. Zhao and Guan presented a unified stochastic and robust UC model by introducing weights for the components of the stochastic and robust parts in the objective function, which was solved by a Benders' decomposition algorithm [20].

This paper explores the optimal DR and UC in a factory's power system, rather than a bulk power system or a DR at home, considering uncertain photovoltaic (PV) power generation. A factory tends to produce more real power from its micro-turbine generators once the time-of-use electricity price set by the utility is high or the PV power is inadequate. The factory has elastic loads, such as production lines, which are associated with interruptible demand and fixed energy consumption in a day. The time periods for operating these elastic loads can be determined by the time-of-use tariff, PV generation, a capacity contract, cost of the fuel of the micro-turbine generator and other operational constraints. The uncertainty of PV power output was addressed by many papers [21–32]. In this paper, the uncertainty of PV power is modeled using stochastic distributions. This problem is expressed by a stochastic mixed integer optimization formulation. The innovations of this paper are summarized as follows:

(1) The problem is solved by a two-level method: the master level determines the optimal states (0/1) of the generators and the elastic loads using a novel genetic algorithm; and the slave level deals with optimal real power scheduling and power purchase/sale from/to the utility, subject to the system operating constraints, using the interior point algorithm.
(2) Two novel encoding schemes associated with new crossover and mutation operations in genetic algorithms are presented. These new operations make this novel GA more efficient to solve the optimal UC and DR in a factory's power system.

(3) The uncertainty in the PV power generation is studied by using the point estimate method that integrates the master level with the slave level to gain an optimal stochastic mixed-integer solution.
(4) Not only the states of micro-turbine generators in UC but also the states of elastic loads at the production lines in DR are addressed at the same time.

The rest of this paper is organized as follows. Section 2 formulates the problem to be solved. Section 3 then presents the methodology based on genetic algorithms. Section 4 summarizes the simulation results of a 30-bus factory power system with PV generations. Section 5 draws conclusions.

2. Problem Formulation

As described in Section 1, the problem can be expressed as follows.

Objective Function

The problem is to minimize the expected value of following objective.

$$f = \sum_{t=1}^{T} \left[\sum_{i=1}^{G_1} \left[u_i(t) \cdot \left(F_i\left(\widetilde{P}_i(t)\right) + S_i(t) \right) \right] \right] + p_p(t) \cdot \widetilde{P}_p(t) - p_s(t) \cdot \widetilde{P}_s(t) \tag{1}$$

where f is the total cost (\$) of power generations from micro-turbine generators plus that of power purchase/sale; T denotes the total number of hours (T = 24 h herein); G_1 represents the number of micro-turbine generators; $u_i(t)$ is the unknown state (0/1) of the i-th unit at hour t; and $F_i\left(\widetilde{P}_i(t)\right)$ and $S_i(t)$ represent the fuel cost (\$/h) and the start-up cost of the i-th unit at hour t, respectively. $\widetilde{P}_i(t)$ and $\widetilde{P}_p(t)/\widetilde{P}_s(t)$ are the unknown real power generated by the i-th unit and the unknown purchased/sold power at the point of common coupling (the swing bus) at hour t. The terms $p_p(t)$ and $p_s(t)$ are the known time-of-use electricity tariffs (purchased/sold power) for this factory. The symbol "~" denotes random variables.

Equality Constraint

At hour t, the objective is subject to

$$\widetilde{P}_p(t) - \widetilde{P}_s(t) + \sum_{i=1}^{G_1} u_i(t) \cdot \widetilde{P}_i(t) + \sum_{j=1}^{G_2} \widetilde{PV}_j(t) = P_{in}(t) + \sum_{\ell=1}^{L} v_\ell(t) P_{el}^\ell(t), \ t = 1, 2, \ldots, T \tag{2}$$

where G_2 is the number of PV arrays; $\widetilde{PV}_j(t)$ is the known power generation from PV array j at hour t, modeled by a stochastic distribution. $v_\ell(t)$ is the unknown state (0/1) of the $P_{el}^\ell(t)$ at hour t. $P_{in}(t)$ and $P_{el}^\ell(t)$ are the known inelastic and known maximum elastic loads (MW) at the ℓ-th production line at hour t, respectively, $\ell = 1, 2, \ldots, L$. If an elastic load is activated ($v_\ell(t) = 1$), then maximum power is consumed (that is, $P_{el}^\ell(t)$); otherwise ($v_\ell(t) = 0$), the power consumption is zero.

Inequality Constraints

At hour t, the powers generated by each micro-turbine generator and each PV array must be within the following limits.

$$P_i^{min} \leq \widetilde{P}_i(t) \leq P_i^{max}, \ if \ u_i(t) = 1, \ i = 1, 2, \ldots, G_1 \tag{3}$$

$$PV_j^{min} \leq \widetilde{PV}_j(t) \leq PV_j^{max} \ j = 1, 2, \ldots, G_2 \tag{4}$$

where P_i^{min} (PV_i^{min}) and P_i^{max} (PV_i^{max}) denote the minimum and maximum of P_i (PV_i). The micro-turbine generators can ramp up and down to higher and lower outputs. The ramp rate (MW/h) is the maximum change in levels (R_i^{up} and R_i^{down}) between two consecutive hours (if the unit is on at time $t - 1$ and t) [33].

$$P_i(t) - P_i(t-1) \leq R_i^{up} \quad (5)$$

$$-R_i^{down} \leq P_i(t) - P_i(t-1) \quad (6)$$

The i-th micro-turbine generator should be operated with a minimum up time M_i^{on} and a minimum down time M_i^{off} [33]. The total elastic load at the ℓ-th production line in a day is fixed because the amount of requested products is firmed.

$$\sum_{t=1}^{T} v_\ell(t) \cdot P_{el}^\ell(t) = P_{total}^\ell \quad \ell = 1, 2, \ldots, L \quad (7)$$

3. Proposed Method

The problem expressed by Equations (1)–(7) consists of $T \times (G_1 + L)$ unknown binary variables and $T \times (2 + G_1)$ unknown random variables. In this paper, $G_1 = 4$, $L = 5$ and $T = 24$. This problem may be solved by binary linear programming or dynamic programming [33–35] in the case that no probability density function is involved. The genetic algorithm (GA), on the other hand, randomly produces many chromosomes, which represent solutions, and selects the fittest one. However, the computational burden of the GA becomes large if the binary bit length of a chromosome and the number of functional constraints is large [36–38]. Thus, this work proposes an enhanced GA to deal with UC and DR.

The proposed method includes two levels of calculations: the master and slave levels. On the master level, two novel methods for encoding generators and elastic demands are presented to overcome the above limitations of the GA and to improve the genetic operations. The GA on the master level solves the problem formulated as Equations (1)–(7) and the constraints of a minimum up time M_i^{on} and a minimum down time M_i^{off} for each micro-turbine generator. When the states of the generators and elastic loads are determined, the remaining problem is to determine $\tilde{P}_i(t)$, $\tilde{P}_p(t)$ and $\tilde{P}_s(t)$ on the slave level, taking into account uncertain PV generation. The interior point method, which incorporates the point-estimation method, is used to solve the problem on the slave level.

3.1. Novel Encoding of Generator States

The traditional GA may encode the on/off states ($u(t)$) of each generator using binary bits (0 or 1), according to Equations (1)–(3). For example, the grey and white parts in a chromosome denote the on and off states, respectively, of a generator in one hour period over 24 h, as shown in Figure 1a.

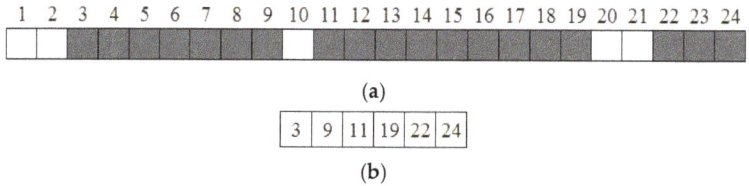

Figure 1. (a) On (grey) and off (white) states of a generator over 24 h; and (b) new encoding.

When the number of generators is large and traditional encoding is used, the length of a chromosome becomes very long. Consequently, this work proposes a pair-based encoding, in which the first and second slots (genes) in a pair represent the starting and ending hours in a period, respectively. Hence, each slot contains integer. Figure 1b shows the new encoding of the chromosome that corresponds to Figure 1a. With this new encoding, the length of a chromosome is much reduced and variable.

Crossover Operation

The crossover operation in this new encoding comprises three steps using a given crossover rate, as follows.

Step A1: Identify the overlapping hours in which states of two selected chromosomes are on, as shown in Figure 2. The proto-offspring becomes (3,4)-(6,9)-(13,19).

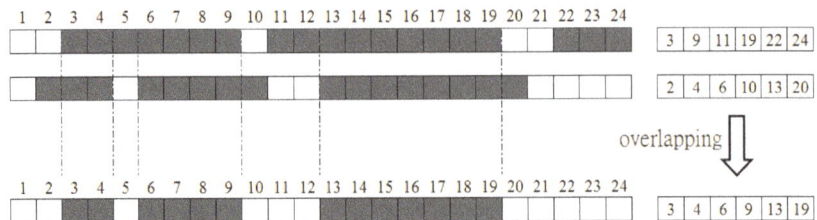

Figure 2. Identification of overlapping hours in which states of both chromosomes are on and resulting proto-offspring.

Step A2: Insert two on-periods arbitrarily. Figure 3a illustrates the insertion of two on-periods. The proto-offspring is (1,1)-(3,4)-(6,9)-(10,11)-(13,19).

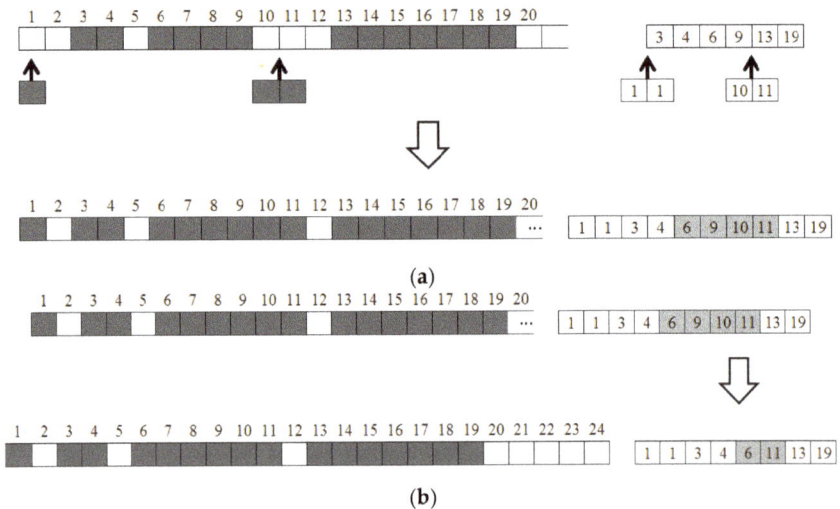

Figure 3. (a) Insertion of two on-periods and resulting proto-offspring; and (b) trimming redundant slot.

Step A3: Trim the redundant slots (genes). Figure 3b reveals that (6,9)-(10,11) can be trimmed to be (6,11). Thus, the final offspring becomes (1,1)-(3,4)-(6,11)-(13,19).

Mutation Operation

A mutation rate for the mutation operation is specified. When a slot is identified, its corresponding integer is mutated to a value that is between those of its neighbors. Figure 4 presents an example: the slot with "11" is identified and mutated to the value of "17", which is between 9 and 19.

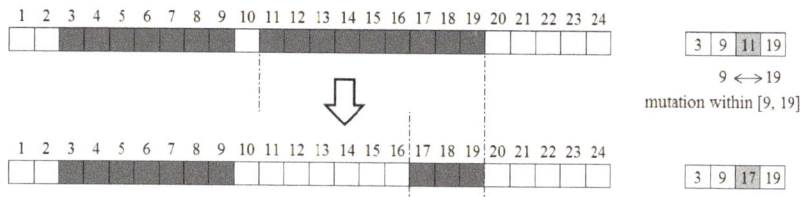

Figure 4. Value of 11 is identified and mutated to 17.

3.2. Novel Encoding of States of Elastic Load

The encoding of elastic loads should differ from that of generators because each elastic load must satisfy the required amount of production daily. This novel encoding is based on the well-known knapsack problem [36].

For example, the grey and white parts in a chromosome denote hours when the state of an elastic load is on or off, respectively, over 24 h, as shown in Figure 5a, using the traditional encoding. The proposed encoding employs several slots (genes) to represent consecutive periods of on, off, on, off and so on states. Restated, the odd and even slots denote the periods where the load is in the on or off state, respectively. Figure 5b shows the results (with respect to Figure 5a) that are obtained by the novel encoding. Notably, the sum of all integers equals 24.

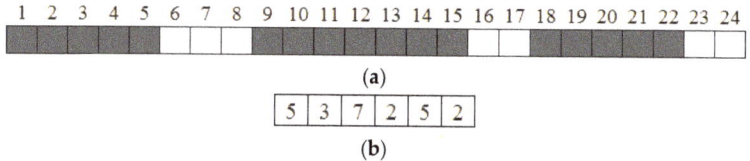

Figure 5. (a) Hourly on (grey) and off (white) states of an elastic load over 24 h; and (b) new encoding corresponding to that in 5a.

Crossover Operation

The crossover operation is performed using the following three steps.

Step B1: Identify a pair of slots (on and off states) in parent 2. The identified pair is inserted in a position between the off and on slots. Figure 6a gives an example in which the total number of hours after insertion is 35.

Step B2: Compute the overvalue (35 − 24 = 11).

Step B3: Trim the integers in the slots to make the total sum 24, as shown in Figure 6b.

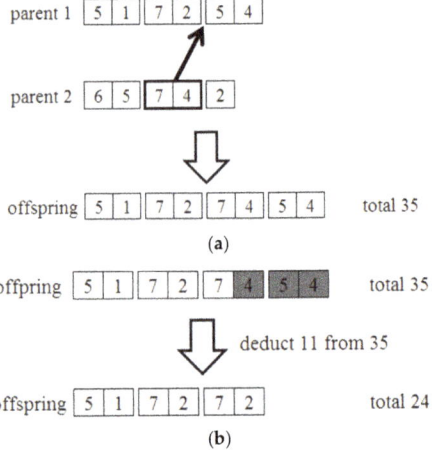

Figure 6. (a) Insertion of a pair of slots; and (b) trimming values in slots to make the sum 24.

Mutation Operation

The mutation operation is carried out by identifying a pair of slots (on and off hours) according to a given mutation rate. The integers in these two slots are mutated to other integers but their sum is unchanged. Figure 7 displays an example in which (7,2) is mutated to (1,8).

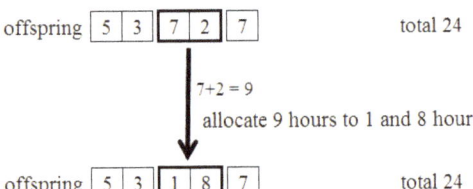

Figure 7. Mutation operation for elastic load encoding.

3.3. Point Estimation Method

Once the states of micro-turbine generators and elastic loads have been determined on the master level, the remaining problem becomes a stochastic optimization problem, which is solved on the slave level. The uncertain PV generation is estimated using the point estimation method [39,40], as follows. Let the expected values, standard deviation and skewness coefficient of each uncertain PV generation (i.e., $\widetilde{PV_j}$) be μ_{PV_j}, σ_{PV_j} and, λ_{PV_j}, respectively, where $j = 1, 2, \ldots, G_2$.

Step C1: Set the first and second moments of the $\widetilde{P_i}(t)$ at the i-th generator, $\widetilde{P_p}(t)$ and $\widetilde{P_s}(t)$ to zero where $i = 1, 2, \ldots, G_1$.

$$E\left(\widetilde{P_i^1}(t)\right) = 0,\ E\left(\widetilde{P_i^2}(t)\right) = 0,\ E\left(\widetilde{P_p^1}(t)\right) = 0,\ E\left(\widetilde{P_p^2}(t)\right) = 0,\ E\left(\widetilde{P_s^1}(t)\right) = 0,\ E\left(\widetilde{P_s^2}(t)\right) = 0 \quad (8)$$

where the symbol E stands for the expectation operator.

Step C2: Compute the two perturbations:

$$\xi_{j,1}(t) = \frac{\lambda_{PV_j}(t)}{2} + \sqrt{G_2 + \left(\frac{\lambda_{PV_j}(t)}{2}\right)^2},\ j = 1, 2, \cdots, G_2;\ t = 1, 2, \ldots, T \quad (9)$$

$$\xi_{j,2}(t) = \frac{\lambda_{PV_j}(t)}{2} - \sqrt{G_2 + \left(\frac{\lambda_{PV_j}(t)}{2}\right)^2}, \ j = 1, 2, \cdots, G_2; \ t = 1, 2, \ldots, T \tag{10}$$

In addition, compute the two weighting factors:

$$\omega_{j,1}(t) = -\frac{\xi_{j,2}(t)}{G_2(\xi_{j,1}(t) - \xi_{j,2}(t))} \tag{11}$$

$$\omega_{j,2}(t) = \frac{\xi_{j,1}(t)}{G_2(\xi_{j,1}(t) - \xi_{j,2}(t))} \tag{12}$$

$$\sum_{j=1}^{G_2} \omega_{j,1} + \omega_{j,2} = 1 \tag{13}$$

Step C3: Estimate the two location parameters:

$$PV_{j,k}(t) = \mu_{PV_j}(t) + \xi_{j,k}(t) \cdot \sigma_{PV_j}(t), \ k = 1,2; \ j = 1,2,\cdots,G_2; \ t = 1,2,\ldots,T \tag{14}$$

Step C4: Let m be 1 where m denotes a moving index given for the PV arrays.
Step C5: Let k be 1. Start to apply the positive perturbation using Equation (9) for the m-th PV array.
Step C6: Find the optimal $P_i(t)$, $P_p(t)$ and $P_s(t)$ where $t = 1, 2, \ldots, T$.

$$\text{Min} f = \sum_{t=1}^{T} \sum_{i \in C1(t)} [F_i(P_{i,mk}(t)) + S_i(t)] + p_p(t) \cdot P_{p,mk}(t) - p_s(t) \cdot P_{s,mk}(t) \tag{15}$$

subject to

$$P_{p,mk}(t) - P_{s,mk}(t) + \sum_{i \in C1(t)} P_{i,mk}(t) + \sum_{j \in G_2, j \neq m} \mu_{PV_j}(t) + PV_{m,k}(t) = P_{in}(t) + \sum_{\ell \in C2(t)} P_{el}^{\ell}(t) \tag{16}$$

$$t = 1, 2, \ldots, T$$

$$P_i^{min} \leq P_{i,mk}(t) \leq P_i^{max}, \ i \in C1(t) \tag{17}$$

and the constraints on ramp rate that are expressed in Equations (5) and (6). The symbols $C1(t)$ and $C2(t)$ are the set of generators and elastic loads in the on state.

Step C7: Let $k = k + 1$. If $k = 2$, then start to conduct the negative perturbation using Equation (10) for the m-th PV array and go to Step C6; else go to Step C8.
Step C8: $m = m + 1$. If $m > G_2$, then go to Step C9; else go to Step C5.
Step C9: Compute the mean and standard deviation of $P_i(t)$, $i = 1, 2, \ldots, G_2$, $t = 1, 2, \ldots, T$.

$$E(P_i(t)) = \sum_{m=1}^{G_2} \sum_{k=1}^{2} \omega_{j,k}(t) \times P_{i,mk}(t) \tag{18}$$

$$\sigma_{P_i(t)} = \sqrt{E(P_i^2(t)) - (E(P_i(t)))^2} \tag{19}$$

The calculations of the means and standard deviations of $\tilde{P}_p(t)$, $\tilde{P}_s(t)$ and cost are similar and omitted here.

Solving the problem in Step C6 becomes easy because it is a quadratic programming problem. This work employs the interior point method in MATLAB to solve this problem [41].

3.4. Overall Flowchart of Algorithmic Steps

The schematic concept of the master–slave iterations can be illustrated as Figure 8. Overall algorithmic steps for the proposed method based on the master–slave iterations are described as follows. Figure 9 illustrates the flowchart of the proposed method. First, specify the elastic loads with their must-run durations and the 24 h inelastic loads; input generator parameters, which are the ramp rate, minimum up-time, minimum down-time and cost coefficients. In addition, the forecasted 24 h PV power generations in terms of means, standard deviations and skewness coefficients are given.

In the proposed GA, both generator units and elastic demands are initialized to be feasible chromosomes. In the master-level calculation, $P_i(t)$ and $P_s(t)$ where $t = 1, 2, \ldots, T$, are determined using Step C1–C9 on the slave level. Please note that if any of the constraints (such as minimum up-time M_i^{on} and minimum down-time M_i^{off} and Equation (7) is violated for a chromosome, its corresponding penalty term is added to the fitness, Equation (1). In the proposed GA, the better chromosomes are selected using the roulette wheel.

As described in the beginning of Section 3, $G_1 = 4$, $L = 5$ and $T = 24$. In a day-ahead scheduling problem, $T = 24$ is always true. Moreover, the number (30 herein) of buses in a factory, which is addressed to consider both UC and DR, is actually large. Considering $G_1 = 4$ and $L = 5$ is reasonable in a large factory power system. This studied problem is not the same as the traditional UC problem, which is defined in the transmission system and may include many buses and generators. Moreover, traditional DR is concerned in the distribution system or at home; however, the DR is emphasized herein in the end-user's factory power system. The proposed GA is very efficient to deal with the operational constraints, such as Equation (7), for the studied problem.

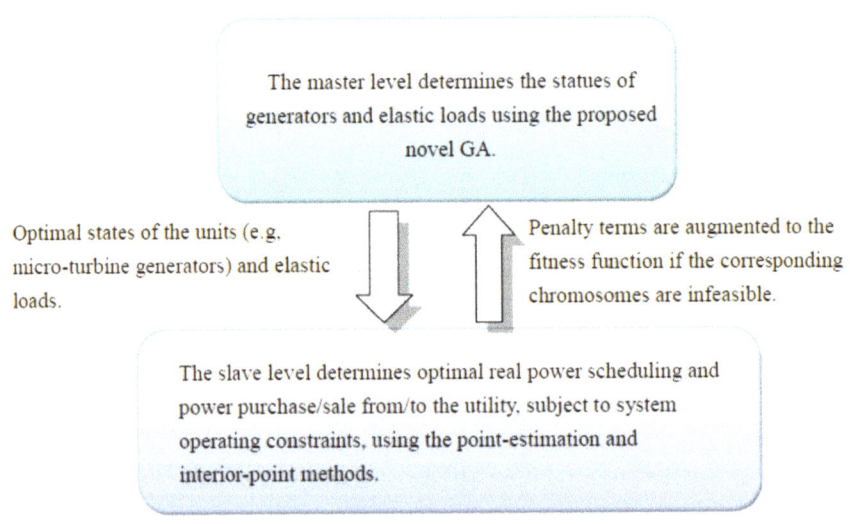

Figure 8. The schematic master–slave iterations.

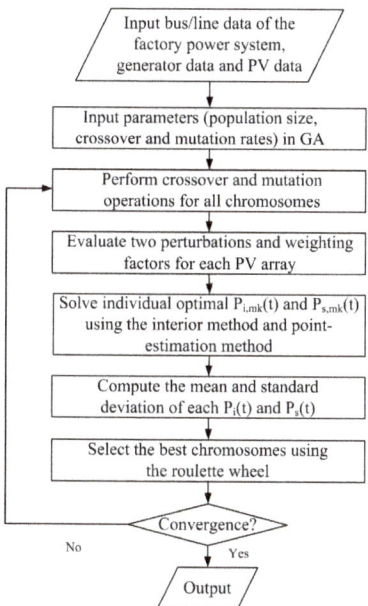

Figure 9. Flowchart of the proposed method.

4. Simulation Results

A 30-bus factory power system, as shown in Figure 10, is used to illustrate the results of the simulation. The micro-turbine generators are at buses 3–6. Table 1 presents the cost coefficients and MW limits of these generators. Table 2 presents the other operational parameters of these generators. The solar PV generations from 7:00 a.m. to 6:00 p.m. at buses 7–10 are modeled using Gaussian distributions, as shown in Figure 11. The standard deviation is set to 3% of the mean value (see page 17 in [42]); that is, the range of PV generation can be estimated approximately to be (the mean value) × (1 ± 3 × 3%), which covers 99.73% of possible PV generation cases. The maximum sizes of these four arrays are 450, 350, 400 and 500 kW. This factory purchases/sells power from/to the utility in a manner determined by the time-of-use prices, as shown in Table 3. In addition to inelastic loads, this factory has five production lines with elastic loads at buses 6, 16, 18, 20 and 26, as shown in Table 4. For example, the total energy consumption of the production line at bus 6 must be 75 MWh (10 × 7.5) daily to produce the required amount of products. Other bus data are given in Appendix A.

A personal computer with Intel (R) Core (TM) i5-2500 CPU @ 3.3 GHZ and 8 GB RAM was used to develop a MATLAB (R2011a, The Mathworks Inc., Natick, MA, USA) code to study the problem. Many scenarios (experimental designs) were studied: PV consideration (see Sections 4.1 and 4.2), different electricity tariffs, M_i^{on} and M_i^{off}, ramp rates, and must-run hours (see Section 4.3). For each scenario, the MATLAB code was run 20 times and the best fitness in these 20 results was considered as the optimal solution. It was found that the proposed novel GA always converges in a finite iterations. Section 4.5 gives the statistics of convergence performance of the proposed method.

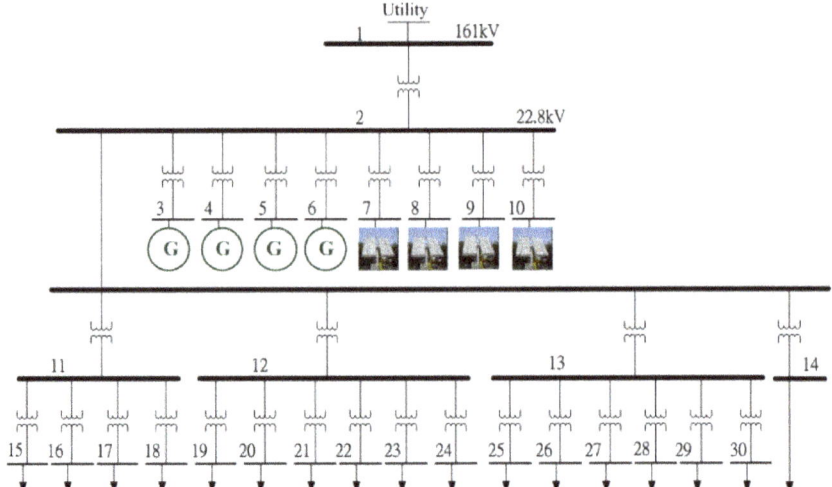

Figure 10. One-line diagram of studied power system.

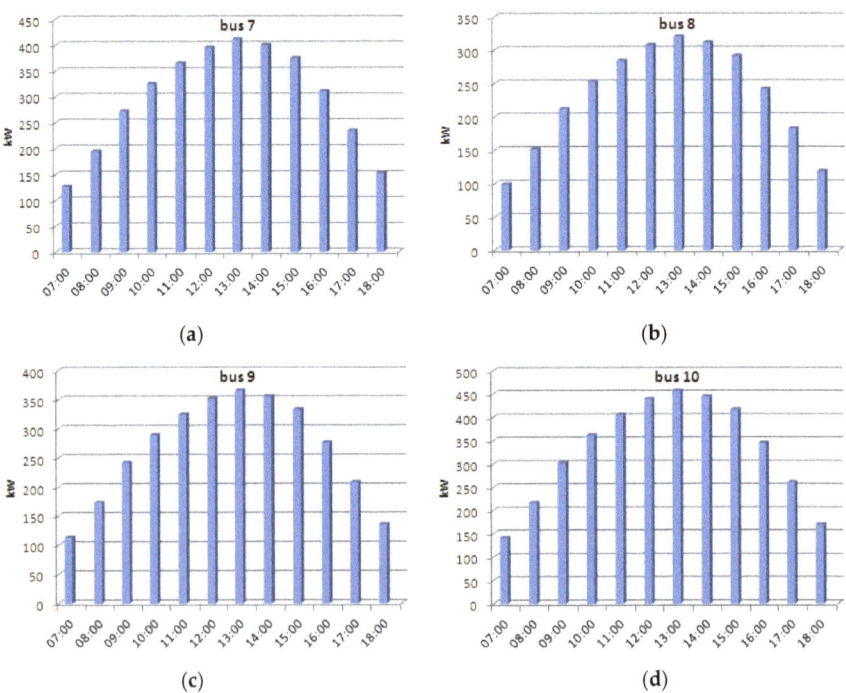

Figure 11. Mean values and standard deviations of PV power at different buses: (**a**) bus 7; (**b**) bus 8; (**c**) bus 9; and (**d**) bus 10.

Table 1. Cost coefficients and MW limits of generators.

Bus No.	P_{min} (MW)	P_{max} (MW)	Cost Coefficients		
			a_i ($/h)	b_i ($/MWh)	c_i ($/MWh^2)
3	0.5	4	151.28	87.87	0.14
4	0.6	3.5	125.21	67.82	0.65
5	0.9	4.5	89.21	31.37	1.1
6	1.1	3	35.48	17.6	0.1416

Table 2. Operational parameters of generators.

Bus No.	Ramp Rate (MW/h)	Start Up Cost ($)	Minimum Up Time, M_i^{on} (h)	Minimum Down Time, M_i^{off} (h)
3	2.5	161.84	4	2
4	2.0	122.53	3	2
5	3.5	175.34	3	1
6	2.0	148.63	4	2

Table 3. Time-of-use pricing.

Periods	Prices ($/MWh)
0:00 a.m.–8:59 a.m.	65
9:00 a.m.–8:59 p.m.	131
9:00 p.m.–11:59 p.m.	65

Table 4. Parameters of elastic loads.

Bus	Must-Run Hours	Maximum Loads (MW)
6	10	7.5
16	6	2.0
18	7	2.5
20	6	2.5
26	5	2.5

4.1. Optimal Solutions Obtained by Proposed Method

The proposed method is applied to find the optimal unit commitment and demand response considering the data described above. The convergence criterion is that 10 consecutive iterations yield the same solution. The studied problem consists of 216 unknown binary variables and 144 unknown random variables.

The optimal solution takes 46 master–slave iterations and 106.99 s to converge. The expected value of total cost is $23,737.68. Table 5 shows the expected values of optimal power injections obtained by the interior point method on the slave level. Table 6 illustrates the expected values of optimal demand responses that are obtained on the master level. The following comments can be made.

(a) A positive expected value of power injection at the swing bus $(\widetilde{P_{sw}}(t))$ indicates that the factory purchases power $(\widetilde{P_p}(t))$ while a negative sign implies that the factory sells power $(-\widetilde{P_s}(t))$ to the utility. The factory sells power to the utility at hours 8:00 a.m.–2:00 p.m., 8:00 p.m. and 9:00 p.m. because the tariff during these periods is high (131$/MWh). In total, 46.1 MWh from the factory is sold to the utility.
(b) Most of the generated PV power is consumed in the factory rather than imported into the utility power system.
(c) Since the tariff for purchasing power from the utility is low during 1:00–8:00 a.m., the elastic loads consume almost all of energy during this period. To fulfill the total demand constraints (75 MWh), the production line at bus 6 also consumes energy during 3:00–7:00 p.m.

Table 5. (a) Expected values of real power injections (MW) during 1:00 a.m.–12:00 p.m. (b) Expected values of real power injections (MW) during 13:00 p.m.–24:00 a.m.

(a)												
t	1	2	3	4	5	6	7	8	9	10	11	12
$P_{sw}(t)$	15.38	13.44	14.19	12.57	12.33	5.31	1.10	−1.62	−6.05	−6.89	−6.41	−6.32
$P_3(t)$	0.50	0.50	0.50	0.50	0.50	0.50	0.50	0.50	3.00	4.00	4.00	4.00
$P_4(t)$	1.15	3.15	2.05	3.50	3.50	1.50	0.60	0.60	2.60	3.50	3.50	3.50
$P_5(t)$	4.50	4.50	4.50	4.50	4.50	4.50	4.50	4.50	4.50	4.50	4.50	4.50
$P_6(t)$	1.59	1.10	1.16	1.10	1.10	0.00	0.00	0.00	0.00	0.00	0.00	0.00

(b)												
t	13	14	15	16	17	18	19	20	21	22	23	24
$P_{sw}(t)$	−6.45	−6.53	1.73	2.04	2.23	2.25	2.27	−5.29	−0.54	5.10	5.01	4.88
$P_3(t)$	4.00	4.00	4.00	4.00	4.00	4.00	4.00	4.00	1.50	0.50	0.50	0.50
$P_4(t)$	3.50	3.50	3.50	3.50	3.50	3.50	3.50	3.50	1.50	1.27	1.17	1.05
$P_5(t)$	4.50	4.50	4.50	4.50	4.50	4.50	4.50	4.50	4.50	0.00	0.00	0.00
$P_6(t)$	0.00	0.00	0.00	0.00	0.00	0.00	0.00	0.00	0.00	0.00	0.00	0.00

Table 6. Expected values of optimal demand response (MW) in a day.

t	1	2	3	4	5	6	7	8–14	15	16	17	18	19	20–24
$P_{el}^6(t)$	7.5	7.5	7.5	7.5	7.5	0.0	0.0	0.0	7.5	7.5	7.5	7.5	7.5	0.0
$P_{el}^{16}(t)$	2.0	2.0	2.0	2.0	2.0	2.0	0.0	0.0	0.0	0.0	0.0	0.0	0.0	0.0
$P_{el}^{18}(t)$	2.5	2.5	2.5	2.5	2.5	2.5	2.5	0.0	0.0	0.0	0.0	0.0	0.0	0.0
$P_{el}^{20}(t)$	2.5	2.5	2.5	2.5	2.5	2.5	0.0	0.0	0.0	0.0	0.0	0.0	0.0	0.0
$P_{el}^{26}(t)$	2.5	2.5	2.5	2.5	2.5	0.0	0.0	0.0	0.0	0.0	0.0	0.0	0.0	0.0

4.2. Optimal Solutions without Considering PV Arrays

This section considers the same condition as considered in Section 4.1 but excluding the PV arrays. In this case, only Step C6 in Section 3.3 was implemented on the slave level. The optimal solution is found in 69 master–slave iterations. The total cost is increased to \$24,333.87 because no PV arrays are utilized to support the demand. Tables 7 and 8 depict the optimal real power injections $P_{sw}(t)$ and demand response during 24 h. The factory sells energy to the utility from 7:00 a.m. to 4:00 p.m. Generally, the power system outside the factory has heavy loads during this period, so the demand response in this factory can mitigate the stress of the utility.

Table 7. (a) Real power injections (MW) during 1:00 a.m.−12:00 p.m. (b) Real power injections (MW) during 13:00 p.m.−24:00 a.m.

(a)												
t	1	2	3	4	5	6	7	8	9	10	11	12
$P_{sw}(t)$	16.37	15.39	14.50	13.67	13.90	1.06	−1.89	−3.25	−3.92	−3.18	−2.92	−2.81
$P_3(t)$	0.50	0.50	0.50	0.50	0.50	1.58	1.08	0.58	1.02	1.51	1.97	2.47
$P_4(t)$	0.06	1.20	1.80	2.40	3.00	3.50	3.48	2.88	3.48	3.50	3.42	3.50
$P_5(t)$	4.50	4.50	4.50	4.50	4.50	4.50	4.50	4.50	4.50	4.50	4.50	4.02
$P_6(t)$	1.14	1.10	1.10	1.10	0.0	1.16	0.0	0.0	0.0	0.0	0.0	0.0

(b)												
t	13	14	15	16	17	18	19	20	21	22	23	24
$P_{sw}(t)$	−4.21	−5.01	−4.35	−4.28	3.12	2.83	2.27	6.21	6.47	−0.51	−0.10	0.25
$P_3(t)$	3.31	4.00	4.00	4.00	4.00	4.00	4.00	0.00	0.00	0.00	0.00	0.00
$P_4(t)$	3.50	3.50	3.50	3.50	3.50	3.50	3.50	3.50	3.48	2.88	2.28	1.68
$P_5(t)$	4.50	4.50	4.50	4.50	4.50	4.50	4.50	4.50	4.50	4.50	4.50	4.50
$P_6(t)$	1.10	1.10	0.00	0.00	0.00	0.00	0.00	0.00	0.00	0.00	0.00	0.00

Table 8. Optimal demand response (MW) in a day.

t	1	2	3	4	5	6	7	8–16	17	18	19	20	21	22–24
$P_{el}^6(t)$	7.5	7.5	7.5	7.5	7.5	0.0	0.0	0.0	7.5	7.5	7.5	7.5	7.5	0.0
$P_{el}^{16}(t)$	2.0	2.0	2.0	2.0	2.0	2.0	0.0	0.0	0.0	0.0	0.0	0.0	0.0	0.0
$P_{el}^{18}(t)$	2.5	2.5	2.5	2.5	2.5	2.5	2.5	0.0	0.0	0.0	0.0	0.0	0.0	0.0
$P_{el}^{20}(t)$	2.5	2.5	2.5	2.5	2.5	2.5	0.0	0.0	0.0	0.0	0.0	0.0	0.0	0.0
$P_{el}^{26}(t)$	2.5	2.5	2.5	2.5	2.5	0.0	0.0	0.0	0.0	0.0	0.0	0.0	0.0	0.0

4.3. Impacts of Different Factors on Total Cost

This section explores the impact of different factors (electricity tariffs, minimum on/minimum down times, ramp rates, and must-run hours in production lines) on the expected value of total cost of the factory, considering the PV generation, with the purpose of validating the proposed method. Table 9 gives the simulation results.

Three time-of-use tariffs are examined. When the tariff during the off-peak hours is high (low), the total cost of the factory is high (low).

When the minimum on-/down-times of generators are all set to 1 h, the constraints are almost relaxed and the acquired cost ($21,879.39) is lower than those in other two cases. When the minimum on/minimum down times are set to large values, these operational constraints are very strict and a larger cost ($25,845.03) is attained.

In the perspective of optimization, large (small) ramp rates imply the problem has a large solution space, resulting in a low (high) cost. Generally, small micro-turbine generators have small ramp rates in factories.

The must-run hours of the production lines reveal the amount of products that must be produced in a day. Long must-run hours, corresponding to many required products, lead to a large total cost, as shown in Table 9.

Table 9. Impacts of different factors on results.

Impact Factors	Descriptions	Expected Values of Total Cost ($)
Electricity Tariffs	Peak hour: 131$/MWh; off-peak hour: 30$/MWh	19,028.80
	As shown in Table 5	23,737.68
	Peak hour: 131$/MWh; off-peak hour: 100$/MWh	25,637.07
Minimum Up/Minimum Down Times (M_i^{on}, M_i^{off})	All are 1 h at all buses	21,879.39
	As shown in Table 2	23,737.68
	(8,4), (8,3), (7,3), (7,4) hours at buses 3, 4, 5, 6.	25,845.03
Ramp Rates ($R_i^{up} = R_i^{down}$, MW/h)	0.5, 0.6, 0.7, 0.8 MW/h at buses 3, 4, 5, 6.	23,781.71
	As shown in Table 2	23,737.68
	4, 3.5, 4.5, 3 MW/h at buses 3, 4, 5, 6.	23,346.11
Must-run Hours in Production Lines	7, 3, 4, 3, 2 h at buses 6, 16, 18, 20, 26	18,254.53
	As shown in Table 6	23,737.68
	15, 11, 12, 11, 10 h at buses 6, 16, 18, 20, 26	33,448.68

4.4. Comparisons between Traditional Method and Proposed Method

The traditional GA (TGA) incorporating with the interior point method and the point-estimation method was used to study the problem for comparisons.

The case studied in Section 4.1 serves as an example here. Table 10 shows the performances of TGA considering different population sizes (50, 75 and 100). The crossover and mutation rates of TGA are 0.8 and 0.02, respectively. The simulation results imply that the TGA yields a feasible solution only because a long binary bit string (24 × 9 = 216 bits) is needed for encoding $u_i(t)$ and $v_\ell(t)$ and the number of real variables (encoded by 144 × 32 bits) is also large. The TGA requires much shorter CPU times and less number of iterations than those needed by the proposed method. Obviously, the TGA converges prematurely. The expected values of final cost obtained by TGA are $31,300.69–31,345.96, which are much greater than $23,737.68 obtained by the proposed method.

Table 10. Performance of traditional GA considering different population sizes.

Population Size	Expected Cost ($)	CPU Time (s)	No. of Iterations
50	31345.96	25.72	15
75	31300.69	54.32	23
100	31321.59	56.07	20

4.5. Statistics of Convergence Performance of the Proposed Method

The simulation results shown in the previous subsections were conducted by running the developed MATLAB code 20 times and the best solution among the 20 results was identified as the final optimal solution. In order to show the convergence performance of the proposed method, this Sections 4.1–4.4 increases the number of runs to 50. The case in Section 4.2 serves as an example.

The best solution occurs in the 9th run of the 50 runs. The best solution is $24,333.87 while the worst one is $30,328.29, which occurs in the 43th run. The difference between $24,333.87 and $30,328.29 are divided equally into 10 portions; that is, 10%, 20%, ... , 100% of the difference between the worst and best costs. As shown in Figure 12a, the probability of the 1st portion is 0.04 (two runs out of 50 runs); the best cost $24,333.87 and the cost, $24,513.49, obtained in the 15th run, are within the 1st portion. The largest probability (0.3) occurs in the 7th portion, which covers the cost between $28,330.11 and $28,996.16. Figure 12b shows the corresponding cumulative probability.

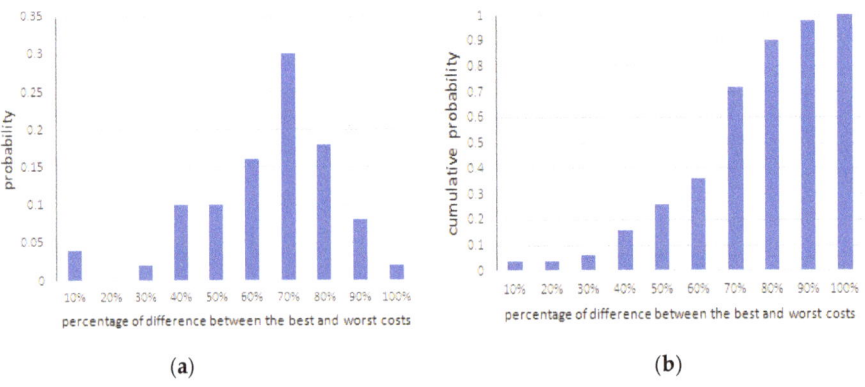

Figure 12. (a) Probability; and (b) cumulative probability with respect to the percentage of difference between the best and worst costs.

4.6. Comparison of Results Considering Different Standard Deviations of PV Power Generations

The standard deviation of PV power generation is set to 3% of the mean value in Section 4.1. That is, the range of PV power generation can be estimated approximately to be (the mean value) × $(1 \pm 3 \times 3\%)$, which covers 99.73% of possible PV generation cases. This section investigates impacts of other standard deviation (8%) on the results. Table 11 shows the comparisons of results by studying the same case in Section 4.1 with different standard deviations of PV power generations. The CPU times and iterations indicate that the proposed GA can still converge properly without premature convergence. The expected cost ($23,028.07) obtained by considering the standard deviation of 8% of the mean value is smaller than that ($23,737.68) obtained in Section 4.1. However, the difference is very small. Actually, the expected power generations of all micro-turbine generators obtained by considering different standard deviations of PV power generations are almost the same because two location parameters of PV power generations in Equation (14) are utilized. These location parameters are centered at the same mean value for a given PV array although different standard deviations are considered and different location parameters are gained in Equation (14).

Table 11. Comparisons of results considering different standard deviations of PV power generations.

Standard Deviation	Expected Cost ($)	CPU Time (s)	No. of Iterations
3%	23,737.68	106.99	46
8%	23,028.07	119.70	46

5. Conclusions

This paper investigates a new problem about optimal UC and DR caused by electricity tariffs in a factory power system. The uncertain amounts of generated power from renewable sources, which may reduce the total cost, are considered in the factory power system. The contributions of this paper can be summarized as follows:

1. The problem concerning optimal DR and UC, considering uncertain PV power generation, in a factory power system, rather than the UC in the bulk power system or DR at home, is formulated and studied.
2. The method based on novel genetic algorithms that are associated with the point estimation and interior point methods is proposed to determine the UC and DR in the factory power system.
3. The proposed string encoding in genetic algorithms efficiently performs both crossover and mutation operations for the UC together with DR. This proposed method ensures that feasible chromosomes can evolve to the fittest solution.
4. Impacts of different parameters (such as PV generations, electricity tariffs, minimum on/down times, ramp rates and must-run hours) were completely investigated on the optimal solutions.

The results of the simulation verify the applicability of the proposed method using a 30-bus factory power system.

Future studies will include modeling different tariffs for electricity and renewable energy as well as different tariffs for purchase and selling energy. Power flow studies in the factory power system will be investigated to ensure that both the voltages and line flows meet the security constraints.

Acknowledgments: The authors acknowledge the financial supported from the Ministry of Science and Technology (Taiwan) under Grant MOST 104-2221-E-033-029.

Author Contributions: Ying-Yi Hong proposed the whole methodology and wrote this article. Po-Sheng Yo developed the program code to realize the method and analyzed the simulation results.

Conflicts of Interest: The authors declare no conflict of interest.

Appendix A

This appendix provides inelastic loads (MW) in the factory power system.

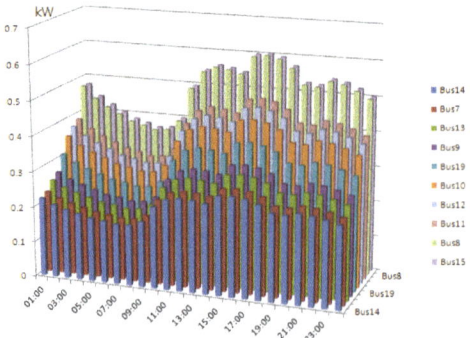

Figure A1. Inelastic loads (MW) in factory power system.

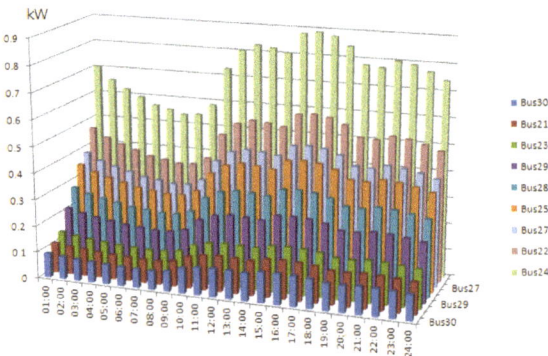

Figure A2. Inelastic loads (MW) in factory power system.

References

1. Bertsimas, D.; Litvinov, E.; Sun, X.A.; Zhao, J.; Zheng, T. Adaptive robust optimization for the security constrained unit commitment problem. *IEEE Trans. Power Syst.* **2013**, *28*, 52–63. [CrossRef]
2. Kalantari, A.; Restrepo, J.F.; Galiana, F.D. Security-constrained unit commitment with uncertain wind generation: The loadability set approach. *IEEE Trans. Power Syst.* **2013**, *28*, 1787–1796. [CrossRef]
3. Bakirtzis, E.A.; Biskas, P.N.; Labridis, D.P.; Bakirtzis, A.G. Multiple time resolution unit commitment for short-term operations scheduling under high renewable penetration. *IEEE Trans. Power Syst.* **2014**, *29*, 149–159. [CrossRef]
4. Federal Energy Regulatory Commission. *Assessment of Demand Response & Advanced Metering: Staff Report*; US Department of Energy: Washington, DC, USA, 2006.
5. Federal Energy Regulatory Commission. *Benefit of Demand Response in Electric Markets and Recommendations for Achieving Them*; U.S. Department of Energy: Washington, DC, USA, 2006.
6. Alexandria, V. Demand Responses II-Building the DSR Road Map. In Proceedings of the PJM Symposium on Demand Responses II-Building the DSR Road Map, Baltimore, MD, USA, 12–13 May 2008.
7. Ott, A.L. Experience with PJM Market Operation, System Design, and Implementation. *IEEE Trans. Power Syst.* **2003**, *18*, 528–534. [CrossRef]
8. Lijesen, M.G. The real-time price elasticity of electricity. *Energy Econ.* **2007**, *29*, 249–258. [CrossRef]
9. Huisman, R.; Huurman, C.; Mahieu, R. Hourly electricity prices in day-ahead markets. *Energy Econ.* **2007**, *29*, 240–248. [CrossRef]

10. Yoon, J.H.; Baldick, R.; Novoselac, A. Dynamic demand response controller based on real-time retail price for residential buildings. *IEEE Trans. Smart Grid* **2014**, *5*, 121–129. [CrossRef]
11. Tsui, K.M.; Chan, S.C. Demand response optimization for smart home scheduling under real-time pricing. *IEEE Trans. Smart Grid* **2012**, *3*, 1812–1821. [CrossRef]
12. Pipattanasomporn, M.; Kuzlu, M.; Rahman, S. An algorithm for intelligent home energy management and demand response analysis. *IEEE Trans. Smart Grid* **2012**, *3*, 2166–2173. [CrossRef]
13. Chavali, P.; Yang, P.; Nehorai, A. A distributed algorithm of appliance scheduling for home energy management system. *IEEE Trans. Smart Grid* **2014**, *5*, 282–290. [CrossRef]
14. Pourmousavi, S.A.; Patrick, S.N.; Nehrir, M.H. Real-time demand response through aggregate electric water heaters for load shifting and balancing wind generation. *IEEE Trans. Smart Grid* **2014**, *5*, 769–778. [CrossRef]
15. Salinas, S.; Li, M.; Li, P. Multi-objective optimal energy consumption scheduling in smart grids. *IEEE Trans. Smart Grid* **2013**, *4*, 341–348. [CrossRef]
16. Tan, Z.; Yang, P.; Nehorai, A. An optimal and distributed demand response strategy with electric vehicles in the smart grid. *IEEE Trans. Smart Grid* **2014**, *5*, 861–869. [CrossRef]
17. Chang-Chien, L.R.; An, L.N.; Lin, T.W.; Lee, W.J. Incorporating demand response with spinning reserve to realize an adaptive frequency restoration plan for system contingencies. *IEEE Trans. Smart Grid* **2012**, *3*, 1145–1153. [CrossRef]
18. Abdollahi, A.; Moghaddam, M.P.; Rashidinejad, M.; Sheikh-El-Eslami, M.K. Investigation of economic and environmental-driven demand response measures incorporating UC. *IEEE Trans. Smart Grid* **2012**, *3*, 12–25. [CrossRef]
19. Zhao, C.Y.; Wang, J.H.; Watson, J.P.; Guan, Y. Multi-stage robust unit commitment considering wind and demand response uncertainties. *IEEE Trans. Power Syst.* **2013**, *28*, 2708–2717. [CrossRef]
20. Zhao, C.Y.; Guan, Y.P. Unified stochastic and robust unit commitment. *IEEE Trans. Power Syst.* **2013**, *28*, 3353–3361. [CrossRef]
21. Kuznetsova, E.; Li, Y.F.; Ruiz, C.; Zio, E. An integrated framework of agent-based modelling and robust optimization for microgrid energy management. *Appl. Energy* **2014**, *129*, 70–88. [CrossRef]
22. Mena, R.; Hennebel, M.; Li, Y.F.; Zio, E. Self-adaptable hierarchical clustering analysis and differential evolution for optimal integration of renewable distributed generation. *Appl. Energy* **2014**, *133*, 388–402. [CrossRef]
23. Hossain, M.J.; Saha, T.K.; Mithulananthan, N.; Pota, H.R. Robust control strategy for PV system integration in distribution systems. *Appl. Energy* **2012**, *99*, 355–362. [CrossRef]
24. Wu, Z.; Tazvinga, H.; Xi, X. Demand side management of photovoltaic-battery hybrid system. *Appl. Energy* **2015**, *148*, 294–304. [CrossRef]
25. Azizipanah-Abarghooee, R.; Niknam, T.; Bina, M.A.; Zare, M. Coordination of combined heat and power-thermal-wind photovoltaic units in economic load dispatch using chance constrained and jointly distributed random variables methods. *Energy* **2015**, *79*, 50–67. [CrossRef]
26. Niknam, T.; Golestaneh, F.; Malekpour, A. Probabilistic energy and operation management of a microgrid containing wind/photovoltaic/fuel cell generation and energy storage devices based on point estimate method and self-adaptive gravitational search algorithm. *Energy* **2012**, *43*, 427–437. [CrossRef]
27. Mohammadi, S.; Mozafari, B.; Solimani, S.; Niknam, T. An adaptive modified fire fly optimisation algorithm based on Hong's point estimate method to optimal operation management in a microgrid with consideration of uncertainties. *Energy* **2013**, *51*, 339–348. [CrossRef]
28. Niknam, T.; Golestaneh, F.; Shafiei, M. Probabilistic energy management of a renewable microgrid with hydrogen storage using self-adaptive charge search algorithm. *Energy* **2013**, *49*, 252–267. [CrossRef]
29. Baziar, A.; Kavousi-Fard, A. Considering uncertainty in the optimal energy management of renewable micro-grids including storage devices. *Renew. Energy* **2013**, *59*, 158–166. [CrossRef]
30. Adi, V.S.K.; Chang, C.T. Development of flexible designs for PVFC hybrid power systems. *Renew. Energy* **2015**, *74*, 176–186. [CrossRef]
31. Alsayed, M.; Cacciato, M.; Scarcella, G.; Scelba, G. Design of hybrid power generation systems based on multi criteria decision analysis. *Sol. Energy* **2014**, *105*, 548–560. [CrossRef]
32. Dufo-López, R.; Bernal-Agustín, J.L. Design and control strategies of PV-Diesel systems using genetic algorithms. *Sol. Energy* **2005**, *79*, 33–46. [CrossRef]

33. Wood, A.J.; Wollenberg, B.F. *Power Generation, Operation and Control*, 2nd ed.; John Wiley & Son, Inc.: New York, NY, USA, 1996.
34. Taha, H.A. *Integer Programming: Theory, Applications, and Computations*; Academic Press: New York, NY, USA, 1975.
35. Swarup, K.S.; Yamashiro, S. Unit commitment solution methodology using genetic algorithm. *IEEE Trans. Power Syst.* **2002**, *17*, 87–91. [CrossRef]
36. Gen, M.; Cheng, R. *Genetic Algorithms and Engineering Design*; John Wiley & Sons: New York, NY, USA, 1997.
37. Suresh, S.; Huang, H.; Kim, H.J. Hybrid real-coded genetic algorithm for data partitioning in multi-round load distribution and scheduling in heterogeneous systems. *Appl. Soft Comput.* **2014**, *24*, 500–510. [CrossRef]
38. Karakatič, S.; Podgorelec, V. A survey of genetic algorithms for solving multi depot vehicle routing problem. *Appl. Soft Comput.* **2015**, *27*, 519–532. [CrossRef]
39. Malekpour, A.R.; Niknam, T. A probabilistic multi-objective daily Volt/Var control at distribution networks including renewable energy sources. *Energy* **2011**, *36*, 3477–3488. [CrossRef]
40. Aien, M.; Fotuhi-Firuzabad, M.; Rashidinejad, M. Probabilistic optimal power flow in correlated hybrid wind–photovoltaic power systems. *IEEE Trans. Smart Grid* **2014**, *5*, 130–138. [CrossRef]
41. The Math Works. *Optimization Toolbox-Fmincon*; MATLAB, The Math Works: Natick, MA, USA, 2009.
42. Ela, E.; Diakov, V.; Ibanez, E.; Heaney, M. *Impacts of Variability and Uncertainty in Solar Photovoltaic Generation at Multiple Timescales*; National Renewable Energy Laboratory: Golden, CO, USA, 2013; TP-5500-58274.

 © 2017 by the authors. Licensee MDPI, Basel, Switzerland. This article is an open access article distributed under the terms and conditions of the Creative Commons Attribution (CC BY) license (http://creativecommons.org/licenses/by/4.0/).

MDPI
St. Alban-Anlage 66
4052 Basel
Switzerland
Tel. +41 61 683 77 34
Fax +41 61 302 89 18
www.mdpi.com

Applied Sciences Editorial Office
E-mail: applsci@mdpi.com
www.mdpi.com/journal/applsci